Where Have All the Heavens Gone?

Where Have All the Heavens Gone?

Galileo's Letter to the Grand Duchess Christina

Edited by

JOHN P. McCARTHY &
EDMONDO F. LUPIERI

CASCADE *Books* • Eugene, Oregon

WHERE HAVE ALL THE HEAVENS GONE?
Galileo's Letter to the Grand Duchess Christina

Cascade Books
An Imprint of Wipf and Stock Publishers
199 W. 8th Ave., Suite 3
Eugene, OR 97401

www.wipfandstock.com

ISBN 13: 978-1-4982-9598-7 (paperback)
ISBN 13: 978-1-4982-9600-7 (hardcover)
ISBN 13: 978-1-4982-9599-4 (ebook)

Cataloging-in-Publication data:

Names: McCarthy, John P., editor | Lupieri, Edmondo F., editor.

Title: Where have all the heavens gone? : Galileo's letter to the Grand Duchess Christina / edited by John P. McCarthy and Edmondo F. Lupieri.

Description: Eugene, OR: Cascade Books, 2017 | Includes bibliographic data.

Identifiers: 978-1-4982-9598-7 (paperback) | 978-1-4982-9600-7 (hardcover) | 978-1-4982-9599-4 (ebook).

Subjects: LCSH: Galilei, Galileo, 1564–1642. | Catholic Church—Italy—Rome—History—17th century. | Religion and science—Italy—History—17th century. | Science—History—17th century.

Classification: QB36.G2 W25 2017 (print) | QB36.G2 (ebook).

Manufactured in the U.S.A.

Contents

Acknowledgments

We want to acknowledge with gratitude the financial and organizational support of the following offices and institutions that made possible the events surrounding the four-hundredth-year commemoration of Galileo's Letter to the Grand Duchess Christina, as well as the production of this volume: the Office of the President, Loyola University Chicago; the Office of the Provost, Loyola University Chicago; the Office of the Dean of the College of Arts and Sciences, Loyola University Chicago; the Office of the Dean of the Graduate School, Loyola University Chicago; the Department of Theology, Loyola University Chicago; the John Cardinal Cody Chair in Catholic Theology, Loyola University Chicago; the Consulate General of Italy, Chicago; the Italian Cultural Institute, Chicago; ItalCultura, Chicago; the Science and Faith–STOQ Foundation, The Vatican; and the Pontifical Council for Culture, The Vatican.

Contributors

George Coyne, SJ, former Director of the Vatican Observatory, currently McDevitt Chair of Religious Philosophy at Le Moyne College in Syracuse, New York

Asim Gangopadhyaya, former chair of the Department of Physics, currently Professor of Physics and Associate Dean for Planning and Resources at Loyola University Chicago

Edmondo Lupieri, John Card. Cody Endowed Chair in Theology and Professor of New Testament and Early Christianity at Loyola University Chicago; President of ItalCultura

Dennis McCarthy, former Director of the Directorate of Time at the Naval Observatory in Washington, DC, and author of *Time: From Earth Rotation to Atomic Physics* (2009)

John McCarthy, former chair of the Theology Department, currently Associate Professor of Systematic Theology at Loyola University Chicago

Mauro Pesce, former Professor of History of Christianity at the University of Bologna and author of *L'ermeneutica biblica di Galileo e le due strade della teologia cristiana* (2005)

Introduction

IF WE USE THE convention of "centuries" to measure human social and cultural history, the seventeenth century in Europe is one of those centuries that is extraordinary. It is the century when the greatest intellectual and artistic labors of figures from Descartes to Shakespeare, from Newton to Rembrandt, from Bach to Cervantes, from Leibniz to Caravaggio took place. It is the century that saw the inventions of the first refracting and first reflecting telescope, the slide rule, the barometer, the pendulum clock, the first human-powered submarine, ice cream, Champagne, and the steam pump. It is the century in which logarithms and calculus came into use. It is the century when Anton van Leeuwenhoek first saw bacteria, when the speed of light was first measured, when the King James Bible was published, when Europe was torn by religious wars, when the first public opera house was opened, when the first newspaper was published, and when the first candy cane was sold.

And it was the century in which Galileo Galilei wrote a letter to the Grand Duchess Christina, the mother of his patron, Cosimo II de' Medici. It would be ludicrous to compare this letter to the works of Shakespeare or Milton, to the intellectual achievements named "calculus" or the microscope, or to the mundane pleasures of Champagne or ice cream. Galileo's letter to the grand duchess is something other than that; it is a classic of a different historical trajectory, the trajectory of the complex interactions of religious and scientific authorities. Within this trajectory, the letter to Christina holds an important place. Written in 1615 but never published until 1636, it was not a monumental publication like *The Message of the Stars* or *The Principia*. In many ways, it was not even original in its science, theology, or its argumentative style. Nonetheless, it is one of those cultural documents that has stood the test of time because in its short span of twenty or so pages it summarizes the struggle of an era, not completely unlike our own, where we both live from the past but hear so clearly that our future is going to be different, unsettlingly different, challengingly different, uncertainly different.

So this is why, four hundred years later, we return to this letter both to remember it and to learn from it. During the academic year that spanned 2015 and 2016, an international group of scholars met at Loyola University Chicago to do exactly that—remember and learn from Galileo's Letter to the Grand Duchess Christina. The year held many events: a production of Bertold Brecht's *Galileo*; a lute and choral concert of the music composed by Galileo's father and brother; a colloquium on Cardinal Bellarmine's announcement that Copernicanism could neither be taught nor held; a symphonic performance of Holst's *Planets*; a demonstration of several of Galileo's original experiments in mechanics; a discussion with Alice Dreger, the author of *Galileo's Middle Finger*, and—what this book records—the bulk of the papers specifically on the text and context of the letter to Christina presented at a two day colloquium.

The first paper is that of the former director of the Vatican Observatory, Fr. George Coyne, SJ. Fr. Coyne has written extensively on the history of science, and particularly on the historical context of Galileo and the controversies of the 1630's that lead to Galileo's trial. In his chapter titled, "Where Have the Heavens Gone? Galileo and the Birth of Modern Science," Fr. Coyne positions Galileo within two contexts, the history of Galileo's life within late fifteenth- or early sixteenth-century Italy, and the history of science and religion at the beginning of the modern era. Fr. Coyne provides the details of the intricate debates about reliable knowledge, about Copernicanism and the Ptolemaic system, about the turn to empiricism and observation in science, about the various writings and publications of Galileo, and about the final trial. As historically rich as this chapter is in detail, it also goes beyond a catalogue of facts to surface what was happening in the changing configuration of pre-modern "science." It is not the case that "science" did not exist in the fifteenth or sixteenth century; it existed as a deductive, largely non-empirical enterprise reliant more on logic than observation or calculation. Fr. Coyne's article explains how with Galileo and others that was changing, not without resistance and misunderstanding all the way around, but changing nonetheless.

John McCarthy's offering, "The Letter to the Grand Duchess Christina (1615): Justice, Reinterpretation and Piety," explores the textual history that led up to the letter of 1615 and focuses on the role that Galileo's appeal to the civic virtue of "piety" played in his arguments about both science and biblical interpretation. Dr. McCarthy, former chair of the Department of Theology at Loyola University, suggests that Galileo was quite intentional in his appeal to piety as a way of advancing a theological position on the "two books" of revelation, an argument that would allow, if understood by the times, a way of preserving the truth of the Bible, the position of the magisterium as articulated by the Council of Trent, and a reliance on methodological observation in the dawning new sciences of the era, without contradiction.

Mauro Pesce is an internationally renowned scholar on Galileo with numerous books and articles on the history, sciences, and events of Galileo's times. In his chapter, "Galileo's Letter to Christina and the Cultural Certainty of the Bible," Prof. Pesce explores the transition in thought that was taking place in Galileo's time from reliable knowledge based on the certainty of tradition, to a whole new form of knowing, one which seemed to many in Galileo's time to be the abandonment of truth altogether. Pesce's contribution focuses on the role of the Bible at the beginning of the sixteenth century for providing unquestionably reliable truth. The Bible was, after all, understood to be the revelation of God, and it was to be interpreted by the magisterium of the Church. Galileo's work was threatening that cultural norm, and that was the source of concern both for the scientific and the religious authorities of the period. Prof. Pesce meticulously explores the texts and intellectual battles that mark this transition in western thought from the Bible as a source of certainty to the more chastened forms of "certainty"—probability, sense experience, methodological observation—characteristic of the modern age.

Dennis McCarthy, the former Director of Time at the United States Naval Observatory and President of the International Astronomical Union Commission on Time, and the Rotation of the Earth, writes on "Galileo's Telescope." Beginning with a summary of the distinctions between physics as natural philosophy and mathematics as a means of establishing numerical relationships, Dr. McCarthy locates the understanding and role of astronomy at the time of Galileo. He then provides a brief history of the reigning conceptions of the planetary configurations, those of Claudius Ptolemy, Nicholaus Copernicus, and Tycho Brahe. The *perspicillum*, later called the telescope, was a relatively new instrument of the times, and when Galileo turns this instrument to the heavens, modern astronomy begins. Dr. McCarthy describes the development of the telescope in Galileo's hands, and what he could have seen with the instruments that he used: the surface of Earth's moon, the moons of Jupiter, Venus, sunspots, and the countless stars that were not visible to the naked eye at night. The heavens for the human mind could never be the same. McCarthy concludes, "In short, the prevailing views of cosmology of the early seventeenth century were threatened. The idea that an appeal to authority could be regarded as a means to explain nature could no longer be accepted in the face of overwhelming evidence."

Asim Gangopadhyaya's chapter continues to explore the science of Galileo's times through a description of some of Galileo's contributions to the field of mechanics. Prof. Gangopadhyaya, former chair of the Physics Department at Loyola University and author of *Supersymmetric Quantum Mechanics*, along with scores of articles on mechanics, describes the experiments that Galileo constructed to understand pendulums, inertia, relativity, uniform acceleration and motion on an inclined plane. Prof. Gangopadhyaya's article illustrates yet another way in which Galileo was turning physics away from a deductive natural philosophy to a science of observation, or experiment and of mathematical measurement.

The seventeenth century was a century of huge cultural shifts and world changing inventions. Among the many extraordinary writings that we have from this century, Galileo's Letter to the Grand Duchess Christina offers a rather intimate record of how the change from a pre-modern to a modern worldview was impacting the life of one of the iconic figures involved in this change. From our vantage point in the twenty-first century, accustomed as we are to cultural plurality and the latest new thing, change in the seventeenth century may seem to occur with glacial speed. And measured by the sheer volume of inventions and the rapidity of cultural movements, Galileo's century is indeed walking, not running. But the letter to Christina provides another measure of change; it documents how difficult it is to change minds, to change institutional habits, to change cultural assumptions that structure the way we act, often without our knowledge. And the pace of those kinds of changes is not all that different today. Galileo's letter remains as an historic and a contemporary testament to the struggle of that kind of change.

John McCarthy and Edmondo Lupieri

1

Where Have the Heavens Gone?

Galileo and the Birth of Modern Science

George V. Coyne, SJ

The Young Professor

On December 7, 1592 the new professor of mathematics at the University of Padua, Galileo Galilei, gave his first lecture in the Great Hall before a large audience. Who was this professor who had not yet turned thirty? Here are a few highlights of his early career which will have a lasting influence on his life as a pioneer of modern science.

While he was attending his second year at the University of Pisa at the beginning of 1583, Galileo had a meeting with the mathematician Ostilio Ricci (1540–1603). At that time Ricci was a tutor in mathematics to the pages of the grand duke of Tuscany, Cosimo I de' Medici, and living in Pisa with the duke's court. From what Galileo asserted towards the end of his life, it was Ricci who introduced him to the study of the geometry of Euclid and afterwards to the mechanics of Archimedes.[1] Thus the fascinating world of clear, rigorous mathematical deduction and of Archimedean science

1. Fantoli, *Galileo*, 40.

was opened to the young Galileo. Here the solution to problems in mechanics was derived not from abstract metaphysical principles but from contact with physical experience. In this way Galileo encountered a method of solving problems which he would go on to perfect and make his own and in so doing contribute to the birth of modern science.

In this regard it is interesting to note that Galileo's earliest preserved letter is precisely the one concerning experiments of his inspired by Archimedes. Galileo sent this letter on January 8, 1588, to Christopher Clavius, the famous Jesuit mathematician of the Roman College,[2] whom Galileo had met while on a trip to Rome in the summer of 1587. Galileo had left Clavius a copy of an Archimedean-inspired theorem of his which formed the basis of his other theorems on barycenters. So there was born at that time a relationship based upon a deep esteem for one another which will remain unfailing right up until the death of Clavius in 1612. It is no wonder that Galileo, still a very young man, was full of admiration for that Jesuit who had by then arrived at the height of his fame as a mathematician.

Through his friendship with Clavius, Galileo developed a wider association with Jesuit philosophers at the Roman College and, in fact, his lectures at the University of Pisa in his early teaching years (1589–1592) were very much influenced by the teachings at the Roman College.[3] From that period as a young teacher Galileo's thinking was taking a direction much more towards natural philosophy than towards pure mathematics. This development will have a profound influence on his slow move towards Copernicanism. In the early 1590s Galileo was deeply impressed upon reading Copernicus and began to consider that the heliocentric theory was superior to the one of Aristotle and Ptolemy. He saw that it was necessary to construct a new theory of motion which was completely detached from that of Aristotle. Thus, for him during his years at Pisa there was a healthy skepticism even with respect to

2. Ibid., 42–43.

3. Wallace, *Galileo and His Sources.*

the material he was teaching, namely Ptolemaic astronomy. He was leaning towards the rival theory of Copernicus.

The opportunity to pursue this thinking came when, at the age of twenty-eight, he began teaching at the University of Padua. He stayed there until 1610, the year when he published his epoch-making telescopic observations. It was a long stay, and Galileo would later write about it as the best period of his whole life. Undoubtedly, this happy experience of Galileo was noticeably influenced by the atmosphere of political and intellectual freedom, of fine manners, and of open hospitality which were characteristic of the Republic of Venice to which Padua belonged. Padua was obviously a happy home for Galileo and it would characterize all of his future confrontations with the Church.[4]

While Galileo publicly taught Ptolemaic astronomy, he was encouraged by the intellectual climate in Padua to pursue his nascent Copernican ideas. But he was aware that he did not yet have sufficient proofs to motivate him to accept Copernicanism unconditionally. Galileo became convinced that there could be only one true system of the world, either that of Aristotle-Ptolemy or that of Copernicus. Throughout his life he would carry on two complementary programs of research: that of a direct proof of Copernicanism and that of constructing a new natural philosophy, opposed to that of Aristotle, which would lay the foundations in physics for Copernicanism and thus open the way for its acceptance. Thus amidst theoretical studies, technical activities, and hours of teaching Galileo passed the first seventeen years of his stay in Padua. But not even in his wildest dreams would he have imagined what the last year of his days in Padua would have in store for him. There was to be a decisive turn of events in his own life and in the history of western thought. It was to bring Galileo indisputable fame, but it would also act as a prologue to the unfolding drama of the second, briefer yet much more intense, part of his existence.

4. Bulferetti, *Galileo Galilei nella società del suo tempo*, 25–28.

The Astronomer and His Telescope

The year 1609 marks a decisive turn of events in the history of astronomy: the beginning of observational astronomy with Galileo's use of the telescope. His observations will be of fundamental importance for eliminating the old view of the world. While he was in Venice, toward the end of July 1609, Galileo received information about the telescope. At that time a stranger arrived in Padua with a telescope which he intended to sell to the Venetian Republic for military use. Galileo decided then and there to make a similar instrument on his own. Upon his return to Padua, he went immediately to work spurred on by the prospect of being able to make a better telescope than the one possessed by the stranger.[5] Galileo went to Venice towards the end of August with an instrument which was quite superior to any made up until that time. It was certainly due to his taking up this technology that the telescope passed from being little more than a curiosity to becoming an instrument of exceptional scientific value for the observations of celestial objects. The first celestial discoveries which occurred in the autumn of 1609 led Galileo to make increasingly powerful telescopes. In November of that same year he succeeded in making one with a magnification power of twenty. Beginning in December he would use this instrument to conduct more accurate observations of the Moon, and in January 1610 he made his most extraordinary discovery of all: the satellites of Jupiter.

The truly revolutionary importance of the results obtained led Galileo to spread the news quickly to the educated community of Europe. And so the *Sidereus Nuncius* (Starry Message) was born. It was published in March of that same year and was written in Latin precisely with a view to its distribution throughout Europe. Galileo was even able to include the very latest observations, made at the beginning of that same month. In this little book of only about sixty pages he first described his observations of the Moon, which proved that the lunar surface was mountainous just like the surface of the Earth. A second sensational discovery concerned the

5. Drake, *Galileo at Work*, 138.

so-called "fixed stars." An enormous number of stars, otherwise invisible to the naked eye, came into view with the telescope. And the Milky Way, whose nature had been debated since antiquity, was seen to be made up of myriads of these stars. The same proved to be the case for certain "nebulae."

But by far the most important observation, according to Galileo himself, was the discovery of four planets which revolved around Jupiter. After providing a detailed report on his observations of them, Galileo concluded that they were of extraordinary importance for his project of constructing a new natural philosophy. For many years he had been persuaded that the Copernican view of the world was by far more probable than the one of Aristotle-Ptolemy.[6] But he was not yet able to justify physically this conviction. Now, however, with his telescopic discoveries, a new and completely unexpected way was opening up for him to search for just such a justification. It was the physical way of sense experience, provided, as a matter of fact, by the new observational instrument. Thus, he was once again pursuing the methodology of Archimedes, that is, a mathematical analysis of sense experience.

These observations allowed him to confirm that on two fundamental points the system of Aristotle-Ptolemy could not be upheld. The first point was that of the essential difference between heavenly bodies, including the Moon, and terrestrial bodies. The existence of mountains and valleys showed that the moon was essentially composed of the same material as the Earth. In fact, by measuring the lengths of shadows he attempted to measure the heights of mountains and the depths of craters on the moon. The second was that of the Earth as the unique center of all celestial motion. The discovery of Jupiter's satellites showed that the motions of heavenly bodies could have centers other than the Earth.

The importance of these discoveries for the Copernican system did not escape Galileo and it was precisely then that he began to conceive the project, as he announced in the *Sidereus Nuncius* itself, of a much more extensive work on a "system of the world," a comprehensive view of the reasons why the system of Copernicus

6. Drake, *Galileo*, 80.

was preferable to that of Aristotle-Ptolemy. This "system of the world" was announced to appear "in a short while," but, as a matter of fact, it would not appear for another twenty-two years as the famous *Dialogue Concerning Two Chief World Systems*.

From then on this project would occupy Galileo's thinking and determine the future course of his life. As he faced the vastness of such a program Galileo considered whether, in order to dedicate himself to it, he should seek an appointment in Florence under the patronage of the grand duke of Tuscany. He had never broken off his relations with Tuscany or with the Medici family. In the preceding years he had gone back regularly in the summertime to Florence as a mathematics teacher to the young prince, Cosimo II de' Medici. This position had given him the opportunity to maintain relationships with the family of the grand duke, Ferdinand I. In order to carry out this program, Galileo asked himself whether he might obtain from the grand duke an assignment which would exempt him from teaching obligations. The negotiations went forward expeditiously and on July 10, 1610, Cosimo II de' Medici appointed Galileo "Principal Mathematician of the University of Pisa and Principal Mathematician and Philosopher to the Grand Duke of Tuscany" without the obligation of residing in Pisa or teaching. It was a lifetime appointment.[7]

Scripture and Copernicanism

Martin Luther's break with Rome in 1519 set the stage for one of the principal controversies to surface in the conflict between the Church and Galileo: the interpretation of Sacred Scripture. In the fourth session of the Council of Trent, the reformation council, the Catholic Church, in opposition to Luther, solemnly declared that Scripture could not be interpreted privately but only by the official Church. As we shall see, Galileo interpreted Sacred Scripture privately; this activity contributed to his condemnation, even though he essentially anticipated by almost 300 years the official teachings

7. Biagioli, *Galileo Courtier*, 159

on the interpretation of Scripture. On November 18, 1893, Pope Leo XIII issued his encyclical *Providentissimus Deus* which called for the study of the languages, literary forms, historical settings, etc. of Scripture so that a fundamentalist approach to Scripture could be avoided. A decade later in 1903 Pius X founded the Pontifical Biblical Institute which is dedicated to just such studies.

One of the first indications that Scripture was to play an important role in the Galileo affair occurred over lunch in 1613 at the palace of the grand duke of Tuscany. The duke's mother, Christina, became alarmed by the possibility that the Scriptures might be contradicted by observations, such as those of Galileo, which seemed to support a sun-centered universe. Since Galileo was supported financially by the grand duke and duchess and in general by the Medici family, this episode was of acute interest to him. Although he was not present, it was reported to him by his friend, Benedetto Castelli. Galileo hastened to write a long letter to Castelli in which he treats of the relationship between science and the Bible. In it Galileo stated what has become a cornerstone of the Catholic Church's teaching:

> I would believe that the authority of Holy Writ had only the aim of persuading men of those articles and propositions which, being necessary for our salvation and overriding all human reason, could not be made credible by any other science, or by other means than the mouth of the Holy Ghost itself. But I do not think it necessary that the same God who has given us our senses, reason, and intelligence wished us to abandon their use, giving us by some other means the information that we could gain through them—and especially in matters of which only a minimal part, and in partial conclusions, is to be read in Scripture.[8]

Galileo was encouraged and supported in his thinking about Scripture by the publication of a letter by the Carmelite theologian, Antonio Foscarini, which favored Copernicanism and introduced detailed principles for the interpretation of Scripture

8. Drake, *Galileo at Work*, 226.

which removed any possible conflict.[9] The renowned Jesuit cardinal, Robert Bellarmine, who would play an important role in the Galileo affair, had already shown an interest in Galileo's telescopic observations. He took the obvious step of consulting the mathematicians among his fellow religious at the Roman College with a letter addressed to them on 19 April. In it Bellarmine posed five precise questions: (1) Was there a multitude of stars invisible to the naked eye? (2) Was Saturn composed of three stars together? (3) Did Venus have phases like the Moon? (4) Was the lunar surface rough and uneven? (5) Did Jupiter have four satellites revolving around it? The reply, signed by Clavius and his colleagues and dated April 24, confirmed the reality of Galileo's discoveries. Nonetheless, Bellarmine responded to Foscarini's arguments by stating that:

> I say that if there were a true demonstration that the sun is at the center of the world and the earth in the third heaven, and that the sun does not circle the earth but the earth circles the sun, then one would have to proceed with great care in explaining the Scriptures that appear contrary; and say rather that we do not understand them than that what is demonstrated is false. But I will not believe that there is such a demonstration, until it is shown me.[10]

However, in the end Bellarmine was convinced that there would never be a demonstration of Copernicanism and that the Scriptures taught an earth-centered universe.

Finally, in June 1615 Galileo completed his masterful Letter to Christina of Lorraine,[11] the same Christina, duchess of Tuscany of the Medici family, who had earlier expressed her concern about the potential contradiction of Scripture. In the letter Galileo essentially proposed what the Catholic Church would, as we have seen, eventually begin to teach. The books of Scripture must be interpreted by scholars according to the literary form, language, and

9. Blackwell, *Galileo, Bellarmine, and the Bible*, 217–51.

10. Finocchiaro, *Galileo Affair*, 67–69.

11. Galilei, *Opere*, 5:309–48.

culture of each book and author. His treatment can be summed up by his statement that

> I heard from an ecclesiastical person in a very eminent position [Cardinal Baronio], namely that the intention of the Holy Spirit is to teach us how one goes to heaven and not how heaven goes.[12]

In the end, however, the Church's Congregation of the Holy Office would declare that placing the sun at the center of the world is "foolish and absurd in philosophy, and formally heretical since it explicitly contradicts in many places the sense of Holy Scripture."[13] The Church had declared that Copernicanism contradicted both Aristotelian natural philosophy and Scripture.

The Dialogue

Let us turn now to the composition of the *Dialogue*. For various reasons the printing of the *Dialogue* took longer than foreseen. Finally, on February 21, 1632, the printer was able to announce its publication and distribution. The *Dialogue* carried the ecclesiastical imprimatur of the vicegerent of Rome, of the master of the Sacred Palace, of the vicar general of Florence, and of the Florentine inquisitor, in addition to that of the government of the grand duke. 1632 was placed as the year of publication. The title had been composed according to the wish of Pope Urban VIII:

Dialogue
of
Galileo Galilei, Lincean
Special Mathematician of the University of Pisa
And Philosopher and Chief Mathematician
of the Most Serene
Grand Duke of Tuscany.

12. Ibid., 5:319.

13. Finocchiaro, *Galileo Affair*, 146; for the original Latin text, see Galilei, *Opere*, 19:321.

Where, in the meetings of four days, there is discussion
concerning the two
Chief Systems of the World,
Ptolemaic and Copernican,
Propounding inconclusively the philosophical and physical reasons
as much for one side as for the other.

According to the customs of those times the text presents a dialogue among three persons: Salviati, who presents the system of Copernicus; Simplicius, who presents that of Ptolemy; and Sagredo, who acts as interlocutor. Although at his trial Galileo will claim that he is presenting reasons "as much for one side as for the other," it is quite obvious that he favors Copernicanism. In the course of the dialogue Galileo puts in the mouth of Simplicius the very ideas that Urban VIII many years earlier had expressed to Galileo, namely that God should not be restricted to having created one system or the other.[14] This move, among others, did not find favor with the pope.

The *Dialogue* is not a dry treatise in astronomy and natural philosophy but rather a polemical writing and, at the same time, a didactic one in support of Copernicanism. In writing it Galileo recalled quite distinctly the long years of battle with the Aristotelians and their unshakable opposition to new ideas and new discoveries. At issue was the entire worldview based on Aristotle's natural philosophy. It had to be demolished so as to prepare the way for the recognition of heliocentrism by educated persons in Italy and all of Europe. But in order to get to that point it was necessary to argue rigorously on the plane of the new natural philosophy.

The *Dialogue* is, therefore, also a work of patient pedagogy in which Galileo seeks with scientific rigor to lead the reader along the road that goes from the old to a new view of the world. Galileo's scientific rigor was based on results already obtained in the field of astronomical observations and on his new natural philosophy, whose foundations he had been laying since his years in Padua. He was now for the first time presenting them to the educated public

14. De Santillana, *Crime of Galileo*, 127.

in expectation of a more systematic and comprehensive formulation which he would publish in 1638 during his final years under house arrest. This last publication is the result of his life-long quest to present a natural philosophy that would correct essential aspects of Aristotle's philosophy. It is known as the *Two New Sciences*,[15] but its official title reveals the momentous character of what is surely Galileo's most significant purely scientific work: *Discourses and Mathematical Demonstrations about Two New Sciences Belonging to Mechanics and Local Motions*. This work does not explicitly refer to Copernicanism; instead it seeks to establish the physics which supports it. With his *Dialogue* Galileo had without a doubt made a decisive contribution to the eventual triumph of Copernicanism. It is ironic, then, that the personal consequence of making this decisive contribution—being sentenced to live the rest of his life under house arrest—provided the opportunity for Galileo to secure in *Two New Sciences* the foundations in natural philosophy for Copernicanism.

The Trial and Condemnation of 1633

As soon as it came off the press the *Dialogue* began to circulate in Italy and Europe, in part because of the numerous copies which Galileo had sent to friends and influential people.[16] The Master of the Sacred Palace, Father Niccolò Riccardi, also received a copy that the inquisitor of Florence sent to him via the Holy Office in Rome. This manner of delivery followed the usual procedure and was not extraordinary. If the *Dialogue* had moved Galileo's friends to praise and admiration, it had also certainly begun to awaken reactions from his opponents. It appears that the opposition had already succeeded in having their voice heard even by Urban VIII. Urban VIII had been made aware of the contents of the *Dialogue* sometime between the end of June and the middle of July and found them disturbing.[17] Urban VIII saw how his arguments

15. Drake, *Galileo at Work*, 373.

16. Westmann, "The Reception of Galileo's Dialogue," 329–37.

17. Shea, "Melchior Inchofer *Tractatus Syllepticus*," 286.

against the possibility of a proof for any world system, made in the course of his lengthy conversations with Galileo years before, had been reduced to one argument only, presented very briefly and, to top that off, by someone named Simplicius. This did not please the pope, and he gave a general order to confiscate all copies of the book in Rome. Riccardi then sought to requisition all of the copies already in circulation. In fact, Riccardi wrote to Florence to advise the Inquisitor that information on the number of copies printed and on their destinations must be diligently gathered so that one could be careful to see that they were had back.

The response of a commission summoned by Urban VIII in August of 1632 only confirmed that Galileo had defended Copernicanism. After five meetings the Commission judged that it was absolutely necessary for the Holy Office to conduct a thorough examination of the *Dialogue*.[18] When he was informed of the commission's position, Urban VIII sent one of his secretaries to the Tuscan ambassador to the Vatican, Francesco Niccolini, so that the latter would give notice of it to the grand duke. After many negotiations among Urban VIII, the grand duke, Father Riccardi, the Holy Office, Niccolini, and others, Galileo appeared before the Commissary of the Holy Office on April 12, 1633.[19]

Through the transcripts of Galileo's trial we are able to know the details of the first interrogation which was held on the same day. After some preliminary questions, the Commissary of the Holy Office, Vincenzo Maculano, began to interrogate Galileo about the events of 1616, the year when Galileo had met with Cardinal Bellarmine at the request of the then reigning pontiff, Paul V. It seems obvious that Maculano wanted to clarify Galileo's responsibility in transgressing the orders he had received from Bellarmine and the Commissary at the time, Father Michelangelo Segizzi. These events will prove to be critical in the final trial and so we must review them.[20]

18. Fantoli, *Galileo*, 286.

19. Ibid., 250.

20. De Santillana, *Crime of Galileo*, 238–41.

In 1616 the Congregation of the Holy Office issued a statement in which Copernicanism was condemned: to hold that "the sun is the center of the world and completely devoid of local motion" was "[therefore] absurd in philosophy [viz., it contradicted Aristotle], and formally heretical."[21] To hold that "the earth . . . moves" received "the same judgment in philosophy."[22] The word "therefore," although not formally in the wording of the decree, is justified and very important. For the consultors of the Holy Office, the natural philosophy of Aristotle was so sacred that to deny it was tantamount to heresy. Soon after the decree appeared, Pope Paul V had Galileo summoned to appear before Cardinal Bellarmine and to accept a private admonition not to promote Copernicanism. In 1633 Galileo was condemned by the same Holy Office for having violated the 1616 injunction by promoting Copernicanism in his *Dialogue*. It must be emphasized that the decree of 1616 did not condemn Galileo (that will occur in 1633). It condemned Copernicanism, which Galileo was suspected of supporting.

Bellarmine had informed Paul V of the ratification of the censure of the two Copernican positions that Galileo held, namely, that the sun stood still in the center of the world, and that the earth moves diurnally, which had taken place at the plenary session of the Consultation of the Holy Office. Both the pope and Bellarmine were convinced that the Copernican thesis was opposed to Scripture. The problem they faced was what to do about Galileo. As Annibale Fantoli writes, "He was by now famous throughout Europe and 'Mathematician and first Philosopher' of the grand duke of Tuscany. Furthermore, there was no doubt about the sincerity of his faith, despite his astronomical ideas. It was probably Bellarmine himself who proposed to Paul V on the occasion of that private meeting with him, the procedure of a private warning. And this would explains the fact that the task to present such a warning was given by the pope precisely to him. With this expedient Galileo

21. Finocchiaro, *Galileo Affair*, 146; for the original Latin text, see Galilei, *Opere*, 19:321.

22. Ibid.

would be silenced once and for all, without offending the grand duke."[23]

Two different documents contained in the trial files verify that Bellarmine carried out his assigned task. They each purport to be reports of the same meetings held on February 25, 1633. The differences between these two documents however are obvious. The second is in perfect agreement with the papal decision of which notice was given to the Holy Office on February 25. Galileo was summoned and subjected to the admonition of Cardinal Bellarmine. The first reports an additional intervention by Segizzi, the Commissary of the Holy Office. But that intervention by the Commissary of the Holy Office was not necessary because such an eventual intervention would have been based on the condition that Galileo would refuse to subject himself to Bellarmine's admonition. Galileo had in fact accepted the private admonition of Bellarmine. Instead, we find in the first document claims that patently contradict the established procedures. In the document, in fact, immediately after Bellarmine's admonition, we find that Commissary Segizzi intervened in a menacing manner. But there is no statement that this intervention was motivated, as prescribed, by Galileo's refusal to accept the admonition. Besides these internal contradictions, there is also the fact, often emphasized in the past, that it lacks the signatures of Bellarmine, Segizzi, the notary who prepared it, and the two witnesses from Bellarmine's household who are named in the document. On Thursday, March 3rd, at the weekly meeting of the cardinals of the Holy Office in the presence of the pope, Bellarmine announced that he had carried out the orders given to him and that Galileo had accepted the admonition. On the other hand, because Segizzi's intervention was untimely and contrary to the instructions given, Bellarmine did not mention it. Segizzi, for his part, was well aware that his intervention was out of order and so he did not dare to add anything to Bellarmine's report. And that explains the content of the document

23. Ibid., 178–79.

relative to this session and the contrast between it and the one dated February 26.[24]

Galileo, of course, wanted to hear what would leak out about the affair in which he was involved so that he might be able to parry the blows that his adversaries would try to direct at him. In fact, rumors began to spread right away that the Inquisition had called Galileo to give an account of his convictions about Copernicanism, that he had abjured these convictions, and that afterwards Cardinal Bellarmine had imposed severe penances on him. Since there was no indication that these rumors were going to die out, Galileo decided to seek recourse with Bellarmine himself. On May 26, Bellarmine issued him a declaration which stated that "Galileo has not abjured in our hands, or in the hands of others here in Rome, or anywhere else that we know, any opinion or doctrine of his; nor has he received any penances, salutary or otherwise."[25]

And so ended an episode in which Galileo and his friends tried to make it appear that there had been no harm to his reputation. But deep down they must have felt an anguishing bitterness. Galileo's plan to silence his opposition by convincing the Church authorities that they must not hastily judge Copernicanism had failed. Despite all of that, Galileo was not the kind to resign himself to defeat. He was convinced that as time went on and with the possibility of finding decisive proofs in favor of Copernicanism still open, the posture of the Church would change. Meanwhile, it was necessary to assume an attitude of prudent reserve and to work quietly to perfect the arguments in favor of heliocentrism. Galileo's adversaries were also not totally satisfied with the conclusion of the affair. The real adversary was not Copernicus, who had been dead for seventy years, but Galileo, who was very much alive and who remained officially uncensored. Of course, something had come out of his having been called before Bellarmine as the gossip about his having abjured Copernicanism reveals. But there were only rumors. An open condemnation of the mathematician of the grand duke would be preferable in order to silence Galileo

24. Fantoli, *Galileo*, 180–82.

25. Baldini and Coyne, *Louvain Lectures*, 25.

definitively. At the same time many of his admirers, sensing a contrary wind, drew back quickly to more secure positions.

Although Galileo produced the declaration of Bellarmine at his trial, it was of no avail and the discrepancy between the two contrasting reports of his private meeting with Bellarmine on February 26, 1616 was never resolved. In the end Galileo was condemned for disobeying the order given to him by Bellarmine. On April 12, 1633, Galileo appeared before the Commissary of the Holy Office. This appearance can be considered the beginning of the trial. Days went by and nothing seemed to happen. But in reality the storm was thickening. In fact, a decision was maturing which was much more rigorous than what could have been hoped for from the information received by Ambassador Niccolini. There was a change in the equilibrium existing up until that time within the Holy Office between a more benign stance toward Galileo and a more rigorous one, which in the end prevailed.[26]

Galileo was not, therefore, prepared for what awaited him when he was requested to present himself to the Holy Office in the morning on Tuesday, June 21. The interrogation began, according to a predetermined plan, with a question by Commissary Maculano that sought to clarify once and for all Galileo's real intention for writing the *Dialogue*. Galileo insisted that he did not hold and had not held the opinion of Copernicus since the command was given to him by Bellarmine in 1616. Father Maculano exhorted him once more to tell the truth because "they will otherwise have recourse to torture." Galileo was then sent back to his dwelling in the Holy Office.[27]

Galileo was kept at the Holy Office until the next day. We may well imagine what must have been his state of mind while he passed those long hours. This was the failure of all of his Copernican efforts, a failure which was by now passing through his mind as inevitable. And, in contrast to 1616, this time his book and he himself would be at the center of the Church's decision. As for the *Dialogue*, Galileo already knew that it would be prohibited.

26. Fantoli, *Galileo*, 323.

27. Ibid., 332.

But what precisely was in store for him? The final illusions which he may have still entertained were completely extinguished when he was forced to put on the penitential garb and was led to the Dominican convent of S. Maria sopra Minerva in the center of Rome where the cardinals and other officials of the Holy Office had gathered together in plenary session.[28]

Galileo was ordered to kneel down and after the reading of the sentence of condemnation he abjured that

> [Since] I wrote and printed a book in which I discuss this new doctrine already condemned and adduce arguments of great cogency in its favor without presenting any solution of these, I have been pronounced by the Holy Office to be vehemently suspected of heresy, that is to say, of having held and believed that the Sun is the center of the world and immovable and that the Earth is not the center and moves: Therefore, desiring to remove from the minds of your Eminences, and of all faithful Christians, this vehement suspicion justly conceived against me, with sincere heart and unfeigned faith I abjure, curse, and detest the aforesaid errors and heresies and generally every other error, heresy, and sect whatsoever contrary to the Holy Church, and I swear that in future I will never again say or assert, verbally or in writing, anything that might furnish occasion for a similar suspicion regarding me.[29]

He was then sentenced to home imprisonment. During this time he completed his great work on the *Two New Sciences* which definitively replaced the natural philosophy of Aristotle and the foundations of the world system of Ptolemy.

28. Ibid.

29. De Santillana, *Crime of Galileo*, 312; for the Italian text, see Galilei, *Opere*, 19:406–7.

Pioneer of Modern Science

The journey of Galileo in his scientific ventures may justly be considered the journey of civilization itself towards modern science. In his early years as a student at the University of Pisa, even before he began his first teaching assignment there, he became acquainted with the methodology of Archimedes which attempts to understand sense experience through the use of mathematics. He pursued this methodology with Christopher Clavius and other Jesuit scientists at the Roman College and then made it a fixture of his whole life as a scientist. It subsequently became a fixture of modern science. The solution to scientific problems was to be derived not from abstract metaphysical principles but from contact with physical experience. Starting with his early years as a teacher, Galileo's thinking was taking a direction much more towards natural philosophy than towards pure mathematics. Mathematics was to become for him the tool with which he would develop models of planetary motion that would challenge those he had inherited. He intuited that it was necessary to construct a new theory of motion which was completely detached from that of Aristotle. This intuition would have a profound influence on his slow move towards Copernicanism. A skeptical approach to inherited ideas would become a hallmark of science.

He was encouraged by the intellectual climate in Padua to pursue his nascent Copernican ideas. From that time on he would carry out throughout his life two complementary programs of research: that of a direct proof of Copernicanism and that of constructing a new natural philosophy, opposed to that of Aristotle, which would lay the foundations in physics for Copernicanism and thus open the way for its acceptance.

Yet another event in Galileo's life would presage the birth of modern science. With Galileo the telescope passed from the level of being little more than a curiosity to becoming an instrument of exceptional scientific value for the observation of celestial objects. From then on science would be inevitably twinned to technology.

With the telescopic discoveries, a new and completely unexpected way was opening up for him to gather scientific data. It was the physical way of sense experience but now the sensor was a new instrument. Taken as a whole his telescopic observations provided the first new significant observational data in over two thousand years and they dramatically overturned the existing view of the universe. They looked to the future. But they threatened the Church.

One of the criteria for the truth value of modern science is its unifying explanatory power, that is, not only are the observations at hand explained but the attempt to understand is also in harmony with everything else that we know, even with what we know from outside the natural sciences.[30] This criterion is significant since it appears to extend the semantics of the natural sciences towards the realm of other disciplines, especially to theology and Christian faith. Put in very simple terms, this criterion is nothing other than a call for the unification of our knowledge. One could hardly be opposed to that. The problem arises with the application of this criterion. When is the unification not truly unifying but rather an adulteration of knowledge obtained by one discipline with the presuppositions inherent in another discipline? History is full of examples of such adulterations. For this reason scientists have always hesitated to make use of this criterion. And yet, if applied cautiously, it could be a very creative one for the advancement of our knowledge and, therefore, of our faith. In this regard Galileo made a significant contribution by confronting the nature of Scripture in his letters to Castelli and to Duchess Christina.

The supposition pursued by Galileo is that there is a universal basis for our understanding and, since that basis cannot be self-contradictory, the understanding we have from one discipline should complement that which we have from all other disciplines. One is most faithful to one's own discipline, be it the natural sciences, the social sciences, philosophy, literature, theology, etc., if one accepts this universal basis. This means in practice that, while remaining faithful to the strict truth criteria of one's own discipline,

30. Coyne, "Destiny of Life and Religious Attitudes," 526–27.

one is open to accept the truth value of the conclusions of other disciplines. And this acceptance must not only be passive, in the sense that one does not deny those conclusions, but also active, in the sense that one integrates those conclusions into the conclusions derived from one's own proper discipline. This was surely what Galileo attempted to do. This, of course, does not mean that there will be no conflict, or even contradiction, between conclusions reached by various disciplines. But if one truly accepts the universal basis I mentioned above, then those conflicts and contradictions must be seen as temporary and apparent. They themselves can serve as a spur to further knowledge, since the attempt to resolve the differences will undoubtedly bring us to a richer unified understanding. Surely, as a pioneer of modern science, Galileo's struggles with the Church have accomplished just that and serve as a spur today to an even richer unified understanding.

Bibliography

Baldini, Ugo, and George V. Coyne. *The Louvain Lectures* (Lectiones Lovanienses) *of Bellarmine and the Autograph Copy of his 1616 Declaration to Galileo*. Studi Galileiani 1, no. 2. Vatican City: Specola vaticana, 1984.

Biagioli, Mario. *Galileo Courtier: The Practice of Science in the Culture of Absolutism*. Chicago: University of Chicago Press, 1993.

Blackwell, Richard J. *Galileo, Bellarmine and the Bible*. Notre Dame: University of Notre Dame Press, 1991.

Bulferetti, Luigi. *Galileo Galilei nella società del suo tempo*. Taranto: Manduria, 1964.

Coyne, George V. "Destiny of Life and Religious Attitudes." In *Life as We Know It*, edited by Joseph Seckbach, 521–534. Dordrecht: Springer, 2006.

De Santillana, Giorgio. *The Crime of Galileo*. Chicago: University of Chicago Press, 1955.

Drake, Stillman. *Galileo at Work: His Scientific Biography*. Chicago: University of Chicago Press, 1978.

Drake, Stillman. *Galileo: Pioneer Scientist*. Toronto: University of Toronto Press, 1990.

Fantoli, Annibale. *Galileo: For Copernicanism and for the Church*. 3rd ed. Vatican City: Vatican Observatory, 2003.

Finocchiaro, Maurice A., ed. and trans. *The Galileo Affair: A Documentary History*. Berkeley: University of California Press, 1989.

Galilei, Galileo. *Le opere di Galilei: Edizione nazionale*. Edited by Antonio Favaro. 20 vols. Florence: Giunti Barbèra, 1890–1909. Reprinted, 1968.

Shea, William R. "Melchior Inchofer *Tractatus Syllepticus*: A Consultor of the Holy Office Answers Galileo." In *Novità Celesti e Crisi del Sapere: Atti del Convegno Internazionale di Studi Galileiani*, edited by Paolo Galluzzi, 283–92. Florence: Giunti Barbèra, 1984.

Wallace, William A. *Galileo and His Sources: The Heritage of the Collegio Romano in Galileo's Science*. Princeton: Princeton University Press, 1984.

Westmann, Robert S. "The Reception of Galileo's Dialogue: A Partial World Census of Extant Copies." In *Novità Celesti e Crisi del Sapere: Atti del Convegno Internazionale di Studi Galileiani*, edited by Paolo Galluzzi, 329–37. Florence: Giunti Barbèra, 1984.

2

The Letter to the Grand Duchess Christina (1615)

Justice, Reinterpretation and Piety

John P. McCarthy

> Aristotle excellently says that we should nowhere be more modest than in discussions about the gods. If we compose ourselves before we enter temples, how much more should we do so when we discuss the constellations, the stars and the nature of the gods, lest from fear or impudence we should make ignorant assertions or knowingly tell lies. (Seneca, *Natural Questions*, Book 7, Chapter 30)

Even on the most cursory reading, it is obvious that Galileo's 1615 letter to the Grand Duchess Christina, the mother of his patron, Cosimo II de' Medici, was written in the midst of several controversies: between Galileo the astronomer and mathematician, and the natural philosophers of the seventeenth-century universities; between Galileo the Catholic and the authority structures of that church; between conservative Aristotelians and reformist Aristotelians; between Protestant reform and Catholic tradition; between the aristocratic values of Florence and the moral universe of Rome.[1] This list of controversies, these struggles of so many "be- tweens," could be extended to include several more, but the point is that any reckoning with the letter to Christina needs to take into account the complex controversies of early seventeenth-century Italy. As a response to controversies of the age, this letter is stead- fastly anchored to a particular time and place.

At the same time this letter seems oddly contemporary for any western reader concerned with the scope of religious author- ity, the role of science in shaping cultural values, the difference between authentic and bogus science and religion, and the con- tributions and limits of scientific thinking for human knowledge. Because this letter is a response to multiple controversies at the intersections of science, religion, politics and culture, issues we deal with today, we can understand this letter as the struggle of a person in an age not completely unlike our own. Anchored as it is in the seventeenth century this letter also has a mooring in our own times.

Located by history in one age and by interest in another, how might we approach the interpretation of this text? It is almost axi- omatic in the field of hermeneutics to recognize that interpreta- tions of a text are, to varying degrees, displays of the interpreter's interests, assumptions, and points of view as well as that of the text. In commenting on the letter to Christina, we must be aware of that. We are not reading this text within a seventeenth-century, north- ern Italian context; we are reading it in the twenty-first century, used to modern science, opposed to religious inquisitions, com- fortable with the privatization of religion, intrigued by the newest and the latest. This is not the culture within which the letter to the grand duchess was written. That was a culture with a deep respect for tradition as the true test of certainty and authority.[2] When we look at this text from the perspective of the twenty-first century

1. Scholarship is unified in locating Galileo's "letter to the Grand Duch- ess Christina" within a context where social, religious, political and academic tensions are evident. For descriptions of this context, see Blackwell, *Galileo*, 29–110; Dietz Moss, "Galileo's Letter to Christina," 547–51; Fantoli, *The Case of Galileo*, 33–96; McMullin, "Galileo on Science and Scripture," 276–89.

2. Pesce, "Galileo's Letter," contained in this volume.

we are likely to see it as an early step toward modern methods of historical and literary biblical interpretation, because that is where biblical interpretation is today. We are likely to read this text as a step toward the increased differentiation of human knowledge and understanding, because we are used to the encyclopedia and to knowledge structured by plural disciplinary methods. We are likely to read this text as an instance in the transition from "certain knowledge" as rigorous deduction from first principles, to the "chastened certainty" of probability, or statistical regularity, because we are on the other side of mathematical sciences and the dethronement of Aristotelian metaphysics. And when we recognize the simple fact that we are twenty-first century readers of a seventeenth-century text one question we might ask is, "what was this text, this particular letter to Christina, about, not as a stepping stone to our age, but as a text for Galileo?"

To be sure, the letter *addresses* biblical interpretation, conflicts about the nature of science, Copernicanism, the new Pythagoreans, issues tied to the Reformation, and larger issues of epistemological and religious transformations. These are issues that Galileo was constantly embroiled in from the time of his education at Pisa in 1581, to his resignation from the faculty of that university after conflicts with the Aristotelians in 1591, to the publication of *Sidereus Nuncius, The Message of the Stars*, in 1610, to the public attacks on his Copernican positions by the "pigeon league" in 1613.[3] The letter to Christina involves or includes all these issues, but as important as these ideas are, something else is at stake. I want to argue in what follows that this letter is not so much about all the individual interpretive controversies as it is about the conditions which call for reinterpretation within traditions that have lost their pliability.[4] Specifically, I want to argue that this letter is about the perception of injustice as a call for reinterpretation, and how Galileo goes about addressing perceived injustice by proceeding with what he calls greater "pious and religious zeal."[5]

3. Fantoli, *The Case of Galileo*, 5–31.

4. Blackwell, *Galileo*, 82.

5. Finocchiaro, "Letter," 113.

Two initial observations are important to get us on the track to the position that I am presenting. First, this text is written as a letter, not as a treatise on science, theology, or the Bible. Neither by intention nor by compositional language (written in Italian and not Latin) was it an academic treatise. It is not filled with the kinds of observations and mathematical constructions that Galileo was quite capable of. Nor was it written as a professional theological argument, something that Galileo was not capable of and knew well enough to avoid. But it did include science, theology, and biblical interpretation written for a more popular audience, the audience of a letter. It is for the most part carefully composed following the rhetorical conventions of the time for a letter, and likely with the assumption that there was going to be some wider readership than Christina alone.[6] Second, it is important to realize that the letter to Christina had a long gestation period, at least back to the 1613 letter that Galileo wrote to the Benedictine monk, Benedetto Castelli.[7] We know from Galileo's writings that he was spending considerable time and effort in the preparation of this letter.[8] Galileo had heard from a friend, Niccolò Arrighetti, that Castelli had been present at a supper conversation with the grand duchess in 1613 where he, Castelli, defended the compatibility of the Bible with the Copernican claim that the sun was the center of our planetary system and the earth, rotating on its axis, revolved around the sun. Present at that supper was Cosimo Boscaglia, a professor of natural philosophy at the University of Pisa, where Galileo had held a position in mathematics. Boscaglia was the first person to accuse Galileo of heresy for defending a Copernican-like position, and thus Galileo interpreted the Castelli—Boscaglia—Christina supper interchange as a proxy discussion of his own views as court philosopher and mathematician in the patronage of Cosimo II de'

6. Throughout this article I am relying on the work of Jean Dietz Moss for the analysis of the letter to Christina in its seventeenth century rhetorical context. Dietz Moss, "Galileo's *Letter to Christina*," 547–76.

7. Ibid, 548–51.

8. Blackwell, "Galileo's Correspondence with Monsignor Dini" in *Galileo*, 216.

Medici, the son of Christina. When Galileo wrote to Fr. Castelli, he praised Castelli for his position, and presented in a calm and measured style most of the arguments regarding biblical interpretation that later appear in the letter to Christina. But we also know from Galileo's subsequent letters to Fr. Piero Dini that he was deliberate in composing the letter to Christina. He did not write this letter with a "fast pen" as had been done with the letter to Castelli.[9] The arguments regarding biblical interpretation were not new; they were, in fact, somewhat commonplace for those who were inclined to the new cosmology. Galileo and others had thought much of that through earlier.[10] But the letter to Christina was not simply a repetition of these gestational writings.

So what do we have thus far? That the letter to Christina was a "popular," non-technical composition, and that the arguments regarding biblical interpretation were in place prior to the 1615 letter. The next step then is to ask: what is significantly different between the 1613 text to Castelli and the 1615 text to Christina? The response to that question will help us get a better understanding of what this particular text to Christina was about for Galileo.

There are three major differences: 1) the tone is markedly different: the letter to Christina is filled with edgy statements, combative accusations, and frustration. You do not find this in the letter to Castelli; there you find most of the major positions on the Bible and science, but with no edge. 2) In the letter to Christina, Galileo has incorporated several references to major church figures, especially to one of Augustine's commentaries on Genesis.[11] Not only is Augustine invoked; so also are Tertullian[12], Jerome[13],

9. Ibid., 216.

10. Blackwell, "Galileo's Letter to Dini" and "Foscarini's Letter to Cardinal Bellarmine" in *Galileo*, 203–7, 217–51.

11. Throughout this article I am using Maurice Finocchiaro's translation of the "Letter to the Grand Duchess Christina." Finocchiaro, *The Essential Galileo*, 109–145. References to Augustine occur on 110, 117, 118, 120, 126, 129–130, 134, 135, 136, and 141.

12. Ibid., 116–117.

13. Ibid., 132.

Dionysius[14], Peter Lombard[15], Thomas Aquinas[16], and the Council of Trent.[17] Galileo knew well that the strength of his position depended on a demonstrated consistency with the more orthodox theological and philosophical traditions of his times. Rational insight and novelty were either not enough, or were suspect in making the kinds of arguments that Galileo had made to Castelli. He needed consistency with major figures in the theological tradition. Castelli helped him, contacting a Barnabite priest, who provided the theological references not found in the 1613 letter to Castelli.[18] 3) Third, and I think most important, is the first quarter of the letter to Christina. In the letter to Castelli, Galileo immediately addresses the issues of biblical interpretation and scientific demonstration. In the letter to Christina, the *captatio benevolentiae* and *narratio*-- standard rhetorical conventions in the structure of letters like this at the time-- are filled with concerns about injustice, imputations of crimes, accusations about heresy, co-opting religion for personal gain, uninformed reading of mathematical texts, hypocrisy, and the like.[19] Galileo knew quite well what the standard interpretation of the pertinent biblical passages pertaining to celestial and terrestrial motion was; he knew quite well how the deductive sciences of his day were practiced; he knew quite well the role of tradition in making the arguments he was making. Some of this, particularly within the practice of deductive science, he was convinced needed to be rethought. This did not entail dismissing the Bible or tradition, or even the goal of certain knowledge, so much a part of Aristotelian science. But in this letter he recognizes that debates about evidence and demonstrations have left the realm of philosophical disputations and become matters of honor, justice and heresy. He understood himself to be the focus of injustice, which, if left unaddressed, would threaten the kind of

14. Ibid., 135, 142.
15. Ibid., 118.
16. Ibid., 132.
17. Ibid., 134.
18. Howell, "Natural Knowledge," 142.
19. Dietz Moss, "Galileo's *Letter*," 551–553.

free inquiry that he believed ought to engage the Bible, science, and tradition. Here is an example of this concern with injustice, one among many in the first part of the letter to Christina:

> Now in matters of religion and reputation I have the greatest regard for how common people judge and view me; so because of the false aspersions my enemies so unjustly try to cast upon me, I have thought it necessary to justify myself by discussing the details of what they produce to detest and to abolish this opinion, in short, to declare it not just false but heretical. They always shield themselves with a simulated religious zeal, and they also try to involve Holy Scripture and to make it somehow subservient to their insincere objectives; against the intention of Scripture and of the Holy Fathers (if I am not mistaken), they want to extend, not to say abuse, its authority, so that even for purely physical conclusions which are not matters of faith one must totally abandon the senses and demonstrative arguments in favor of any scriptural passage whose apparent words may contain a different indication.[20]

There are at least two dimensions of injustice that Galileo is concerned with in this text and in the letter as a whole. The first dimension consists of the public and private attacks against him by individuals of the conservative "pigeon league": Lodovico delle Colombe, Niccolò Lorini, and Tommaso Caccini.[21] This dimension of the perceived injustice we might call the "proximate dimension". It is this dimension that causes Galileo to begin his letter to Christina with his concerns about rumors, injuries, and the imputation of crimes, falsehoods and heresies. But there is a second, related dimension of injustice that Galileo feels compelled to address throughout the entire letter. We might label this dimension of perceived injustice the "structural dimension." This is the kind of perceived injustice that Galileo associates with "the impious," with "pretend religion," with "hypocritical zeal," arrogance, misuse of biblical and ecclesial authority, unwillingness to understand, and

20. Finocchiaro, "Letter," 113.

21. Fantoli, *Case of Galileo*, 52–53; 61; 72–75; 93–96.

insincerity. It is this injustice that Galileo argues will ultimately allow "the infidel" to make a laughing stock of the Christian tradition because it will have locked the Church into designations of truth, untruth, and heresy about issues far exceeding the scope of the rightly ordered understanding of scripture and the goal of religion, namely, worship of God and salvation of souls.[22] It is a structure that has taken hold of both the Church and the university which systematically distorts the true goals and practices of these institutions.

Galileo does not hesitate to use the language of justice, injustice and crimes to speak about the proximate dimension of justice; he does not use this same language to speak about the structural dimensions of justice that constitute the rationale for most of his arguments about the many conflicts that he addresses throughout the letter. So more must be said about this structural dimension of perceived injustice.

For Galileo the structural dimension of injustice is not a criticism of the very structures of the Christian tradition or the Roman Church as the Protestant criticisms of the time were. Rather the structural dimension of perceived injustice was a criticism of a disordered use of these structures by individuals, schools, and habits of thought directed toward ends other than that proper to Christianity itself. It was, in Aristotelian terms, a telic criticism. An institution should act with a view to the good, whether it be the family, the government, or the Church. And the good to which the Church is directed is the worship of God and the salvation of souls in Galileo's view. Injustice occurs not only when individuals lie or spread rumors (proximate injustice) but when the institution becomes complicit with the arrogance of individuals, even enabling it. In doing so the institution loses sight of its own good and in the process supports injustice in the name of a justly ordered institution. Galileo thought the Church was losing sight of its proper role in rightly directing humanity to God. For example, Galileo writes:

22. Finocchiaro, "Letter," 138.

> Of this sort are those who try to argue that this author should be condemned without examining him; and to show that this is not only legitimate but a good thing, they use the authority of Scripture, of experts in sacred theology and of sacred Councils. I feel reverence for these authorities and hold them supreme, so that I should consider it most reckless to want to contradict them when they are used in accordance with the purpose of the Holy Church; similarly I do not think it wrong to speak out when it seems that someone, out of personal interest, wants to use them in a different way from the holiest intention of the Holy Church.[23]

In the particular controversies that Galileo was involved in at the time, the right ordering of the Church involved integrating interpretive principles and practices in three areas: scripture, nature and logic. The perceived structural injustice that concerned Galileo was the result of not following, or not properly integrating, these principles and practices because of the various forms of "impiety" that he experienced: arrogance, misuse of authority, etc. In Galileo's view, piety itself had become corrupted. The rightly ordered relationship of the individual to the institution of the Church, and the institution of the Church to God, was seriously threatened. Galileo's remedy for injustice is "to proceed with much more pious and religious zeal than they."[24]

Galileo's intention to proceed with much greater piety sounds strange to twenty-first century ears because we hear "piety" on the other side of romanticism, Pietism, and the privatization of religion. It is not a feeling that Galileo is appealing to; it is not a renewed sense of private devotion, or a new Gnostic religious allegiance. Galileo's solution to the injustice he sees is an appeal to a traditional public civic virtue of his time, and the times from the Roman rhetoricians through Augustine to the works of the high Middle Ages, including Thomas and beyond—Piety.[25] Piety as a

23. Ibid., 114.

24. Ibid., 113.

25. Garrison, *Pietas: From Vergil to Dryden*; Broadie, "Aristotelian Piety," 54–70; McPherran, "Piety, Justice, and the Unity of Virtue," 299–328.

public virtue was the right ordering of the relation between God, state, and family, all of which one is responsible to and for, but all of which must be kept in right order. If they are not kept in right order then two things happen—duty is not carried out, or, if it is carried out, it is ordered to the wrong end or done in an imbalanced way. And this is what Galileo, in the *petitio* of the letter[26] seeks—to keep the right order and balance—give to sense experience and demonstration its due; give to the Bible its due; give to tradition its due; give to authority its due. Because of misplaced piety, he reasons there is a false zealousness all around that makes the *conventional interpretations* of the Bible, science, and tradition, *destructive misinterpretations* registered as injustice.

Much more than biblical interpretation is at stake when Galileo is invoking piety. He is arguing from a theological position that both the Bible and nature proceed from God as a revelation of God. Each of these two avenues of revelation has its own unique form of interpretation, and because each proceeds from the same God, they must be consistent. What proceeds from God through the inspiration of the Holy Spirit are words accommodated to a particular age and capable of expressing mysteries far exceeding the capacities of human reason to understand their meanings. What proceeds from God in nature is the direct execution of the will of God; it is the effect of the command of God in creation. Thus nature is established, a structure to be understood by close sense observation rather than textual interpretation. Monotheism is the theological foundation for the logical truth that these two approaches to the revelation of God cannot be in contradiction. Thus piety, respect for the right order of the individual, the institution and God, requires the respect for, and recognition of, the proper means for understanding the revelation of God through these two avenues. "Greater piety" respects the Bible, nature and logic. Thus Galileo writes:

> Therefore I think that in disputes about natural phenomena one must begin not with the authority of scriptural passages, but with sense experiences and necessary

26. Dietz Moss, "Galileo's *Letter*," 553–57.

> demonstrations. For the Holy Scripture and Nature derive equally from the Godhead, the former as the dictation of the Holy Spirit and the latter as the most obedient executrix of God's orders; moreover to accommodate the understanding of the common people it is appropriate for Scripture to say many things that are different (in appearance and in regard to the literal meaning of the words) from the absolute truth; on the other hand, nature is inexorable and immutable, never violates the terms of the laws imposed upon her, and does not care whether or not her recondite reasons and ways of operating are disclosed to human understanding; but not every scriptural assertion is bound to obligations as severe as every natural phenomenon; finally, God reveals Himself to us no less excellently in the effects of nature than in the sacred words of Scripture.[27]

For Galileo what is behind the proximate and structural injustice that he opposes are various forms of impiety ranging from false accusations to disregard of God's revelation. And the measure of the right response to this is to advance a greater "pious and religious zeal" or a "sincere religion" which respects the right order of the individual, the institution and God through the respect of the proper understanding of the Bible, nature and logic.

Just as the letter to Christina is not a treatise on science or theology, so also it is not a treatise on the relation of piety and justice. Rather it attempts to redress perceived injustice through reinterpreting a distorted piety, a false zealousness, by demonstrating, in practice, the complexity of a greater "pious and religious zeal." Understanding this greater "pious and religious zeal" then means clarifying how the respect for nature, logic, and for the proper understanding of the Bible work together to form, in Galileo's thinking, the way to respect God's revelation.

The historian and philosopher of science, Erin McMullin, has offered a helpful way to clarify the more theoretical sides of the practice of Galileo when Galileo himself is not offering a treatise on theology, science, or biblical interpretation. McMullin

27. Finocchiaro, "Letter," 116.

recognizes Galileo's consistent reliance on Augustine's, *De Genesi ad litteram*, and observes that Galileo was utilizing the same principles for interpreting God's revelation in scripture and nature that Augustine had used a thousand years earlier.[28] McMullin provides slightly differing lists of these principles in various articles[29], but the overall perspective governing both Augustine's and Galileo's thinking includes, in McMullin's analysis, these five principles:

> 1) The Principle of Consistency (PC): The proper meaning of scripture cannot be in true conflict with the findings of human sense or reason.[30]
>
> 2) The Principle of Priority of Demonstration (PPD): When there is a conflict between a proven truth about the physical world and a particular reading of scripture, an alternative reading should be sought.[31]
>
> 3) The Principle of the Priority of Faith (PPF): An apparent demonstration on the side of philosophy of something that is contrary to a doctrine of faith must be set aside.[32]
>
> 4) The Principle of Accommodation (PA): The choice of language in the scriptural writings is necessarily accommodated to the capacities of the intended audience.[33]
>
> 5) The Principle of Scriptural Limitation (PSL): Since the primary concern of scripture is with human salvation, we should not look to scripture for knowledge of the natural world.[34]

28. McMullin, "Galileo on Science and Scripture," 292.

29. McMullin, "Galileo on Science and Scripture," 292–99; "Galileo's Theological Venture," 197–203.

30. McMullin, "Galileo's Theological Venture," 197.

31. Ibid., 197.

32. Ibid., 198.

33. Ibid., 199.

34. Ibid., 199.

McMullin is perceptive in the distillation of these Augustinian principles and correct to observe Galileo's reliance, in practice, on these principles when dealing with the two avenues of revelation. To be sure, these principles are part of Galileo's greater "pious and religious zeal," but I would venture to say that a greater "pious and religious zeal" for Galileo, dealing as it does with the Bible, nature and logic, is even more intricate than what McMullin helpfully identifies through these five principles of interpretation. In the letter to Christina, Galileo is not about the task of providing a handbook of piety; he is about presenting the case that his Copernican position is pious and Catholic in the face of injustice. Thus the principles that are at work in presenting this case are operative rather than explicitly defined. Because these principles are operative rather than explicit in form, it may be best to see the principle in action, in the words of the letter, rather than through commentary. Thus in what follows I offer what I see to be the principles, fifteen in all, that animate Galileo's intention to proceed against injustice with a greater piety.

In the area of scriptural interpretation there are, I would suggest, at least seven operative principles.

1. Principle of True Meaning:

    > It is most pious to say and most prudent to take for granted that Holy Scripture can never lie, as long as its true meaning has been grasped.[35]

2. Principle of the Right End: [when speaking of accommodation to popular understanding]

    > This would be especially implausible when mentioning features of these created things which are very remote from popular understanding, and which are not at all pertinent to the primary purpose of the Holy Writ, that is, to the worship of God and the salvation of souls.[36]

35. Finocchiaro, "Letter," 115.

36. Ibid., 116.

3. Principle of Right Authority:

    That is, I would say that the authority of the Holy Scripture aims chiefly at persuading men about those articles and propositions which, surpassing all human reason, could not be discovered by scientific research or by any other means than through the mouth of the Holy Spirit himself.[37]

4. Principle of Accommodation:

    Since these propositions dictated by the Holy Spirit were expressed by the sacred writers in such a way as to accommodate the capacities of the very unrefined and undisciplined masses, therefore for those who deserve to rise above the common people it is necessary that wise interpreters formulate the true meaning and indicate the specific reasons why it is expressed in such words.[38]

5. Principle of Right Wisdom:

    . . . in the learned books of worldly authors are contained some propositions about nature which are truly demonstrated and others which are simply taught; in regard to the former, the task of wise theologians is to show that they are not contrary to Holy Scripture; as for the latter (which are taught but not demonstrated with necessity), if they contain anything contrary to the Holy Writ, then they must be considered to be indubitably false and must be demonstrated such by every possible means.[39]

6. Principle of Coherence with Tradition:

    From this and other places it seems to me, if I am not mistaken, the intention of the Holy Fathers is that in questions about natural phenomena which do not involve articles of faith one must first consider whether they are demonstrated with certainty or known by sense

37. Ibid., 117.
38. Ibid., 116.
39. Ibid., 126.

> experience, or whether it is possible to have such knowledge. When one is in possession of this, since it too is a gift from God, one must apply it to the investigation of the true meaning of the Holy Writ at those places which apparently seem to read differently.[40]

7. Principle of Right Scope:

> For I do not see that in this regard they [Councils] prohibit anything but tampering, in ways contrary to the interpretation of the Holy Church or of the collective consensus of the Fathers, with those propositions which are articles of faith, or which involve morals and pertain to edification according to Christian doctrine: so speaks the Fourth Session of the Council of Trent.[41]

In the area of interpreting nature, I would suggest there are at least four operative principles:

1. Principle of the Book of Nature:

> Therefore I think that in disputes about natural phenomena one must begin not with the authority of scriptural passages, but with sense experiences and necessary demonstrations. For the Holy Scripture and Nature derive equally from the Godhead, the former as the dictation of the Holy Spirit and the latter as the most obedient executrix of God's orders.[42]

2. Principle of Sense Experience:

> Nature is inexorable and immutable, never violates the terms of the laws imposed upon her, and does not care whether or not her recondite reasons and ways of operating are disclosed to human understanding . . . and so it seems that a natural phenomenon which is placed before our eyes by sense experience or proved by necessary demonstrations should not be called into question,

40. Ibid., 130.
41. Ibid., 134.
42. Ibid., 116.

> let alone condemned, on account of scriptural passages whose words appear to have a different meaning.[43]

3. Principle of Disciplined Observation:

> Here it should be noticed, first, that some physical propositions are of a type such that by any human speculation and reasoning one can only attain a probable opinion and a verisimilar conjecture about them, rather than a certain and demonstrated science; an example is whether the stars are animate. Others are of a type such that either one has or one may firmly believe that it is possible to have, complete certainty on the basis of experiments, long observations, and necessary demonstrations . . . I should think, as stated above, that it would be proper to ascertain the facts first, so that they could guide us in finding the true meaning of scripture.[44]

4. Principle of Differentiated Knowledge:

> I think no one will say that geometry, astronomy, music and medicine are treated more excellently and more exactly in the sacred books than in Archimedes, Ptolemy, Boethius, and Galen. So it seems that the royal preeminence belongs to it [theology] in the second sense, namely, because of the eminence of the topic, and because of the admirable teaching of divine revelation in conclusions which could not be learned by men in any other way, and which concern chiefly the gaining of eternal bliss.[45]

In the area of logic, I would suggest that there are, as with nature, at least four operative principles:

1. Principle of Non-Contradiction:

> Because of this, and because (as we said above) two truths cannot contradict one another, the task of the

43. Ibid., 116–117.
44. Ibid., 128–29.
45. Ibid., 124–5.

wise interpreter is to strive to fathom the true meaning of the sacred texts; this will undoubtedly agree with those physical conclusions of which we are already certain and sure through clear observations or necessary demonstrations.[46]

2. Principle of Propositions and Words:

Returning to the preceding argument, if we keep in mind the primary aim of the Holy Writ, I do not think that its always saying the same thing should make us disregard this rule; for if to accommodate popular understanding Scripture finds it necessary once to express a proposition with words whose meaning differs from the essence of the proposition, why should it not follow the same practice for the same reason every time it has to say the same thing?[47]

3. Principle of Prudent Reasoning [Galileo quoting Augustine]:

Now then, always practicing a pious and serious moderation, we ought not to believe anything lightly about an obscure subject, lest we reject (out of love for our error) something which later may be truly shown not to be in any way contrary to the holy books of either the Old or New Testament.[48]

4. Principle of Rational Responsibility:

However, I do not think one has to believe that the same God who has given us senses, language, and intellect would want to set aside the use of these and give us by other means the information we can acquire with them, so that we would deny our senses and reason even in the case of those physical conclusions which are placed

46. Ibid., 120.

47. Ibid., 130–31.

48. Ibid., 110.

> before our eyes and intellect by our sense experiences or by necessary demonstrations.[49]

Two comments are in order here after suggesting these operative principles in Galileo's greater "pious and religious zeal." First, as with any interpretive practice, these operative principles do not constitute a rigorous method for Galileo, but rather a way of thinking which guides the selection of reliable authorities, the choice of values, and the placement of emphasis within contested interpretations. As much as Galileo is presenting a new method for science which relies on sense experience and necessary demonstration, he is also presenting a method for the practice of piety, one that was structured less on obedience to authority and more on the recognition of the right ordering of the individual, the institution, and God within a dynamic tradition. Second, for all its artificiality and abstraction from Galileo's very specific concerns and arguments, this list of operative principles is an indication of the complex and interrelated patterns of thought that Galileo was bringing to the conflicts he had to face. No one of these principles dictated Galileo's response. More like a racing scull with multiple oars pulling together to move the boat forward, Galileo constantly, and often not explicitly, utilized interpretive skills at the intersection of several fields of thought simultaneously. As seems often the case, the unity involved in the practice of thought like that displayed in Galileo's letter defies the attempt to adequately understand or easily explain this same unity.

Galileo was not a theologian and he consistently admitted that. But he was sensitive to the theological issue at stake, and to the consequences that either misplaced or misguided theological thinking could generate. Galileo agreed with Augustine's advice to keep "always our respect for moderation in grave piety." Whether Galileo read further in Augustine we do not know for sure, but given the schooling that he had, it is not unlikely. And, for sure, we will never know if he read Augustine's *Enchiridion* or *De doctrina*, works designed, like Galileo's letter, to "get interpretation right." It

49. Ibid., 117.

seems likely that he would agree, however, with what Augustine wrote to Laurentius, at the beginning of that *Enchiridion*:

> The true wisdom of man is piety. You find this in the book of holy Job. For we read there what wisdom itself has said to man: "Behold the fear of the lord [*pietas*], that is wisdom." If you ask further what is meant in that place by *pietas*, the Greek calls it more definitely θεοσέβεια, that is worship of God. The Greeks sometimes call piety εὐσέβεια, which signifies right worship. Though this, of course, refers specifically to the worship of God. But when we are defining in what man's true wisdom consists, the most convenient word to use is that which distinctly expresses fear of God.[50]

In the four hundred years since Galileo wrote his letter to Christina, the understanding of piety has changed significantly. Like religion, it has been privatized, dismissed as a civic virtue and put aside as the topic that one would want to use in order to explore how to integrate new knowledge with tried traditions in a cultural context. So it seems strange to us, a kind of odd historical curiosity of the period, that Galileo would invoke the language of piety when dealing with the dawning age of modern science. But it is important to recognize, I think, that Galileo is not being "pious" in our modern sense of the term; he is pious in a much more robust way, a way quite consistent with the history of piety as a civic virtue of right order. He is dealing with injustice, proximate and structural, by appeal to this sense of piety as right order and with the hope that this kind of right order could be seen not, as a threat to tradition, but as the way that tradition itself could live dynamically. In this sense, Galileo was a pious person of his age.

Bibiography

Augustine, *Enchiridion*. Translated by J. F. Shaw. South Bend: Regnery/Gateway, 1961.

Broadie, Sarah. "Aristotelian Piety." *Phronesis* 48 (2003) 54–70.

50. Augustine, *Enchiridion*, 2.

Blackwell, Richard J. *Galileo, Bellarmine and The Bible*. Notre Dame: Notre Dame University Press, 1991.

Dietz Moss, Jean. "Galileo's *Letter to Christina*: Some Rhetorical Considerations." *Renaissance Quarterly* 36 (1983) 547–76.

Fantoli, Annibale. *The Case of Galileo: A Closed Question?* Translated by George V. Coyne SJ. Notre Dame: Notre Dame University Press, 2012.

Finocchiaro, Maurice A. *The Essential Galileo*. Indianapolis: Hacket, 2008.

Garrison, Jane D. *Pietas from Vergil to Dryden*. University Park: Pennsylvania University Press, 1987.

Howell, Kenneth J. "Natural Knowledge and Textual Meaning in Augustine's Interpretation of Genesis: The Three Functions of Natural Philosophy." In *Nature and Scripture in the Abrahamic Religions*, edited by Jitse M. van der Meer and Scott Mandelbrote, 117–146. 2 vols. in 4. Brill's Series in Church History 36-37. Leiden: Brill, 2008.

McMullin, Ernan. "Galileo on Science and Scripture." In *The Cambridge Companion to Galileo*, edited by Peter Machamer, 271–347. Cambridge Companions to Philosophy. Cambridge: Cambridge University Press, 1998.

McMullin, Ernan. "Galileo's Theological Venture." *Zygon* 41 (2013) 192–220.

McPherran, Mark L. "Piety, Justice, and the Unity of Virtue." *Journal of the History of Philosophy* 38 (2000) 299–328.

Pesce, Mauro. "Galileo's Letter to Christina and The Cultural Certainty of the Bible." Vontained in this volume.

3

Galileo's Letter to Christina and the Cultural Certainty of the Bible

Mauro Pesce

The unabated influence of the biblical hermeneutics that Galileo presents in his letter to Christina[1] on the history of modern intellectual life over the last four centuries has been described many times.[2] In several essays published thirty years ago, I tried to reconstruct the stages of the letter's redaction and reception.[3] The first draft is represented by Galileo's letter to Benedetto Castelli (December 21, 1613), which Galileo enlarged and modified many times until the last versions were finished in May-June 1615.[4] The letter to Christina circulated in Europe first in manuscript form and then in the published Strasbourg edition of 1636. It was the center of a wide-ranging debate that took place in France for several decades after Galileo's trial and condemnation. It also made an important contribution to the debate on the impact of

1. The letter is published in Galilei, *Opere*, ed. Favaro, 5:309–48. The most accurate study of the letter is Damanti, *Libertas philosophandi*. See also Pesce, "La «Lettera a Cristina»," 7–66; and Pesce, "Galileo a Cristina," 35–84.

2. Pesce, "L'interpretazione della Bibbia," 239–84; Pesce, "Momenti della ricezione," 55–103; and Pesce, "Il Consensus Veritatis," 53–76.

3. Pesce, *L'ermeneutica biblica di Galileo*, 117–95.

4. Ibid., 29–85 and see n29 below; Pesce, "Una nuova versione della lettera," 89–122; Pesce, "Le redazioni originali della lettera," 394–417.

Copernicanism on modern Christian theology.[5] Seventeenth-century Catholic theologians constantly discussed it (see, for example, the *Almagestum Novum*, Bologna 1651, by the Jesuit Giovanni Battista Riccioli). Vincenzo Ferrone has shown that in that same period Galilean biblical hermeneutics continued to constitute an important point of reference in Italian environments that decisively denied any opposition between Christianity and modern science. Ferrone mentions, for example, scientists such as Giuseppe Valletta, Francesco D'Andrea, Lucantonio Porzio, Costantino Grimaldi and Eustachio Manfredi.[6] The republication of the letter to Christina in 1710 helped to boost Galileo's ideas on the Bible. Galileo's hermeneutics therefore received renewed attention, for example, by Ludovico Antonio Muratori in his *De ingeniorum moderatione* (Paris, 1714) and by Pietro Giannone (in 1734).[7] William Derham's famous *Astro-Theology* (London, 1714) became another vehicle in the debate on Galilean biblical hermeneutics, a debate which continued to develop through the nineteenth century until Galilean hermeneutics finally received a timid reception in Leo XIII's encyclical *Providentissimus Deus*.[8]

The hermeneutical theories of Galileo, alongside his friends' and adversaries' responses to them, are today the object of a complex and endless discussion.[9] See, for example, the *vexata quaestio* of the so-called "concordism"[10] in the interpretation of the Bible and in the last part of the letter to Christina. Another *vexata quaestio* concerns Roberto Bellarmine's proposal to consider Copernicanism a hypothesis, a proposal that Galileo refused in some of his

5. Pesce, "Il Copernicanesimo e la teologia," 33–46.

6. Ferrone, *Scienza natura religione*, 3–16.

7. Giannone, "Parere intorno la censura del padre Massimiliano Galler gesuita," 367–413.

8. Pesce, *L'ermeneutica biblica di Galileo*, 160–73.

9. Pesce, "L'indisciplinabilità del metodo," 151–74; Pesce, "Gli ingegni senza limiti," 637–59.

10. Regarding the question of concordism, see Pesce, *L'ermeneutica biblica di Galileo*, 224–29.

"Copernican" writings.[11] Other questions concern the declaration of heliocentrism as heretical by a commission of theologians appointed by the Holy Office (February 26, 1616) and eventually the trial of Galileo himself in 1633.

The opening of the Archive of the Congregation of the Doctrine of the Faith on the case of Galilei in 1998 was an important event that, however, as Adriano Prosperi has affirmed, "has not changed much in our knowledge."[12] It is, however, true that the opening of the archive has resulted in new editions of the acts of the trial.[13] The studies produced by the Pontifical Commission appointed by Pope John Paul II to review the so-called Galileo case resulted mostly in apologetic defenses[14] of the theological positions taken by Catholic institutions over the past centuries and have failed to convince the international community of historians of the modern age and science.[15] Nevertheless, two renowned Catholic scholars, George Coyne and Annibale Fantoli, have produced studies that are the opposite of those of the Pontifical commission.[16]

11. Ibid., 117–73.

12. Prosperi, "L'Inquisizione e Galilei," 18.

13. Pagano and Luciani, *I documenti del processo*; Pagano, *I documenti vaticani del processo*; Beretta, *Galilée devant le tribunal de l'inquisition*; Beretta, "Le procès de Galilée et les Archives du Saint-Office." See also Finocchiaro, *The Trial of Galileo*.

14. Regarding the distinction between history and apologetics in recent studies on Galileo's trial, see Prosperi, "L'Inquisizione e Galilei," 9–30.

15. See, e.g., Poupard, "La controversia tolemaico-copernicana," 15–21; and Brandmüller, *Galilei e la chiesa*. Paolo Galluzzi rightly criticizes the historical reliability of the apologetic historiographical thesis according to which (1) Galileo would not have understood that in that phase of history Copernicanism was just a hypothesis, (2) some theologians of the time, but not Cardinal Bellarmine, would not understand how the Scriptures had to be properly interpreted, and (3) as soon as it was provided with unquestionable evidence of the physical reality of the heliocentric thesis, the Church hastened to accept this vision implicitly acknowledging that its condemnation was a mistake (Galluzzi, "Il 'caso Galileo,'" 6).

16. Coyne, "The Church's Most Recent Attempt," 340–59; Fantoli, "Galileo e il Vaticano," 273–97 (in this article Fantoli criticizes Artigas and Sánchez de Toca, *Galileo e il Vaticano*); Id., "Il caso Galileo," 229–57; for Fantoli's chief

All these questions have been extensively debated in the last decades and I have expressed my opinions about them many times in the past.[17] For this reason, in the following pages I want to face the interpretation of the letter to Christina and Galilean biblical hermeneutics from a rather new point of view.

I. The Letter to Christina: The Bible is No Longer the Main Basis of Cultural Certainty

In every culture there are many bases of certainty. They constitute the undisputed cultural assumptions that underpin the ability to live everyday life. Human thinking and acting presupposes an immense complex of cultural data which are ordinarily considered absolutely certain. It is certain that the sun rises and sets, that the life is divided between waking and sleeping. It is certain that a good seed will bear fruit under the right climatic conditions. It is certain that the language one uses is understood by members of his or her family and associates. For every human being who belongs to a particular culture there are multiple deposits of cultural certainties. They are, first of all, (a) the common conceptions about nature and human life; (b) practices of daily life, ways of reacting to the surrounding world, mechanisms for organizing space and time; (c) values and customary norms; and (d) complexes of traditional and religious knowledge, sacred texts and places, and objects and images considered supernatural. All that seems culturally certain is never questioned by those who belong to a particular group. In fact, in order to be able to question and replace what is culturally certain, one must possess an alternative base of certainty.

Of course, in every culture, there are debates and conflicts concerning the proper interpretation that ought to be given to the cognitive, ethical, practical, and religious bases that make up the certainty that underpins social life. In traditional society, as defined by Talcott Parsons, however, dissent is justified essentially

work, see his *Galileo*.

17. Pesce, *L'ermeneutica biblica di Galileo*, 82–213; Pesce, "Il Copernicanesimo e la teologia," 33–46.

in two ways. First, dissent is expressed through *interpretations* of the cultural base that diverge from institutional or commonly accepted interpretations. However, since the cultural base is essentially religious in traditional societies, a second way to legitimize dissent consists in attempting to directly reach the ultimate source of traditional cultural certainty by bypassing the mediation of commonly accepted institutions. It is this direct access (for example, through heavenly revelations) that can legitimize the claim of certain subjects to possess the authentic interpretation of the traditional religious base. Supernatural revelations, however, do not consist in critical investigation but derive from the presumed direct access to an authority considered unquestionable. To summarize, interpretation and recourse to a presumed direct contact with the source of religious cultural certainty are the two forms that can legitimate dissent in traditional societies.

Unlike traditional societies, modern societies have come to possess an alternative source of knowledge, completely independent of the traditional cultural certainties. This new source exists because of the birth of modern science. In the letter to Christina Galileo compares the fundamental base of cultural certainty of his time, that is, the Bible, with the experimental scientific method which is entirely independent of any traditional base of certainty.

In Galileo's society the Bible was one of the foundations of cultural certainty. It was not, however, an isolated cultural artifact. It was rather deeply connected to the whole symbolic system of Christian societies. After the Christianization of the Aristotelian system (that took place beginning in the thirteenth century),[18] the overall conception of the universe was the result of a merger between biblical conceptions and the Ptolemaic system. The earth is located at the center of the universe and stands still. Hell is placed in its depths. The house of God and paradise are found above the heavens. Heliocentrism, therefore, does not just contradict biblical statements about the mobility of the sun and the stability of the earth. It also implies the impossibility of the traditional localizations of hell and paradise. It makes Jesus's ascent into heaven and

18. Gregory, "Filosofia e teologia nella crisi del XIII secolo," 61–76.

descent into hell impossible. The theologians of Galileo's time were well aware of these problems. Explicit statements to this effect were made, for example, by Tommaso Campanella in his *Apologia pro Galileo* in 1616[19] and by Melchior Inchofer, who, in his *Tractatus Sylleptícus* (1633), was one of the theologians arguing for the necessity of condemning Galileo.[20]

What moves Galileo in his research is the discovery of new knowledge: "Who will set bounds to human ingenuity? Who is going to claim that everything in the world which is observable and knowable had already been seen and discovered?"[21] For Galileo the starting point of scientific research is not methodic doubt or even radical skepticism about the traditional certainties. The tradition is an integral part of his academic and economic life practice. The questioning of opinions "which are almost universally accepted" is only the consequence of scientific conclusions that have demonstrated their inconsistency. In any case, the effect of the new discoveries was precisely to show the falsity of common opinions that constituted the cultural sources of certainty: "Nor should it be considered rash to be dissatisfied with opinions which are almost universally accepted."[22] Hence the need for an absolutely certain knowledge:[23]

> I say that the human intellect does understand some of them [propositions] perfectly, and thus in these it has as much absolute certainty [*certezza*] as nature itself has. Of such are the mathematical sciences alone; that is, geometry and arithmetic, in which the divine intellect indeed knows infinitely more propositions, since it knows all. But with regard to those few which the human intellect

19. Campanella's *Apologia pro Galileo* was published in Frankfurt in 1622. For an Italian translation, see Fomiano, *Apologia per Galileo*.

20. Inchofer, *Tractatus syllepticus*, 31–32. See also Pesce, *L'ermeneutica biblica di Galileo*, 13. 102. 145–8; and Beretta, "'Omnibus Christianae, Catholicaeque Philosophiae amantibus,'" 301–27.

21. Galileo to Christina, trans. Finocchiaro, *The Galileo Affair*, 96–7, with my corrections.

22. Ibid., 97.

23. "Certainty" is a word that appear more than 160 times in Galileo's work.

> does understand, I believe that its knowledge equals the divine in objective certainty [*certezza*], for here it succeeds in understanding necessity, beyond which there can be no greater sureness.[24]

This need is characteristic of most of the philosophers and scientists of the early seventeenth century. Consider, for example, Descartes's *Meditationes de prima philosophia*:

> Some years ago I was struck by the large number of falsehoods that I had accepted as true in my childhood, and by the highly doubtful nature of the whole edifice that I had subsequently based on them. I realized that it was necessary, once in the course of my life, to demolish everything completely and start again right from the foundations if I wanted to establish anything at all in the sciences that was stable and likely to last [*si quid aliquando firmum et mansurum cupiam in scientiis stabilire*]. . . . Reason now leads me to think that I should hold back my assent from opinions which are not completely certain and indubitable just as carefully as I do from those which are patently false.[25]

For Descartes, it is necessary to subject all the traditional certainties to a radical methodical doubt "in order that there may at length remain nothing but what is certain and indubitable" (*quod certum est et inconcussum*).[26] The need of Descartes, like that of many intellectuals and scientists in the first half of the seventeenth century, was the pursuit of a knowledge that can be "entirely certain and indubitable."

This was not the attitude of Galileo, as we said. Galileo did not intend to reform the culture of his time. He was a scientist, not a cultural or religious reformer. His purpose was to devote himself to the knowledge of nature in its different aspects, from physics

24. Galilei, *Dialogue Concerning the Two Chief World Systems*, trans. Drake, 103. For the Italian, see Galilei, *Dialogo sopra i due massimi sistemi*, in *Opere*, ed. Favaro, 7:128–9.

25. From the first meditation, in Descartes, *Meditations on First Philosophy*, ed. and trans. Cottingham, 22–23.

26. From the second meditation, in ibid., 35.

to astronomy. His statement, quoted above, "Who will set bounds to human ingenuity? Who is going to claim that everything in the world which is observable and knowable had already been seen and discovered?"[27] best expresses his purpose. Thus in a way that differed from the path of Descartes, which was theoretical and philosophical, Galileo came to find the certainty to which Descartes and many others would later aspire.

Galileo believed that he had found the basis for absolutely certain and indubitable knowledge. This knowledge was offered by the experimental method, which analyzes nature on the basis of two converging factors: the sense experiences (*sensate esperienze*) and the necessary demonstrations (*necessarie dimostrazioni*). Sense experience is not based on the five senses which the human body possesses, but on scientific instruments that enable the correction of what is perceived through the five senses. "Necessary demonstrations" are necessary precisely because they rely on mathematical procedures. Thus the conclusions reached by the scientific method are, for Galileo, absolutely certain. Robert S. Westman could affirm that, "In his defense of the Copernican system Galileo committed himself to a strict notion of proof in science, according to which true conclusions must be deduced necessarily from true premises, which are themselves self-evident. No other Copernican had locked himself into such a tight position."[28] The scientific method therefore does not rely on any data that comes from common or traditional opinions. It is not based on data received from the ancient cultural knowledge shared by the members of a society. The scientific method is the only indubitable means to certainty available to humanity, with the exception of that particular field that exceeds human capacity to know and that is defined by the Catholic faith.

27. Galileo to Christina, trans. Finocchiaro, *The Galileo Affair*, 96–97.

28. Westman, "The Copernicans and the Churches," 98.

II. The Theory Proposed by Galileo in the Letter to Christina

In 1609[29] Galileo began to use the telescope.[30] In March 1610 he published his astronomical observations in the *Sidereus Nuncius*. As is known, his Copernican ideas aroused opposition in some ecclesiastical circles. The Dominicans of Florence, Niccolò Lorini and Tommaso Caccini, battled relentlessly against him from 1612 to 1615 in an attempt to condemn the scientist for heresy. They argued that the astronomical theories of Galileo were contrary to the assertions of the Bible on the mobility of the sun and the stability of the earth. A second type of opposition came from Aristotelian academic environments and was linked to scientific and theological positions. Over the centuries, the Ptolemaic astronomical system was Christianized and became the basis for an understanding of the cosmos which represented the sacred cosmological vision of Christian theology. Consequently, the defense of the Aristotelian-Ptolemaic worldview was also the defense of the Christian one.

Galileo's astronomical discoveries and his Copernicanism therefore contradicted not only Ptolemaic astronomical theories but also threw into crisis a comprehensive theological system that was deeply enmeshed in the Ptolemaic view of the universe. Not surprisingly, the reactions against Galileo were both scientific and theological.

Galileo was therefore obliged to fight on two fronts; one way he did so was to separate theological arguments from scientific ones. Facing theological criticisms derived from the Bible, Galileo worked out, between 1613 and 1615, a theological theory that argued against the use of the Bible in matters of natural science. The most complete formulation of this theory is in the long letter to Christina of Lorraine. The final drafting of the letter is perhaps in May 1615. But later revisions are not to be excluded.[31]

29. Pesce, "L'indisciplinabilità del metodo," 162; see also Guthke, who speaks about *Bewusstseinveränderung* in *Last Frontier*.

30. Van Helden, *Invention of the Telescope*.

31. The final text of the letter to Christine is the result of a long series of

The central implication of Galileo's theoretical proposal is that the experimental scientific method of modern science creates a new epistemological foundation. Galileo did not intend to criticize the Christian faith. He rather proposed a new configuration of the cultural bases of his society which opened the possibility for a new and fundamental distinction between science and religion.

Galileo's theoretical proposal was methodological. Scientific knowledge about nature stems from the experimental scientific method, namely from "sense experiences" and "necessary demonstrations". For Galileo, "scientific experiment" and mathematical analysis produce absolutely certain cognitive conclusions that cannot be changed or adapted—not even by the scientist who produced them:[32]

---

revisions that begins on December 21, 1613, with the letter that Galileo wrote to his disciple, Benedetto Castelli. The Dominican friar Niccolò Lorini sent a copy of this letter to the Holy Office on February 7, 1615. Lorini's aim, which he did not attain, was to get Galileo condemned for heresy. In the period between February and May 1615 Galileo wrote the letter to Christina using his letter to Castelli as a base. A first phase of transforming the letter to Castelli into the letter to Christina is located in a recension of the former that I have found in a Latin translation in the appendix to Pierre Gassendi's *Apologia in Io. Bap. Morini librum* (1649) (see Pesce, *L'ermeneutica biblica di Galileo*, 55–85). In this version we see clearly some extensions of the letter to Castelli that we encounter later in the letter to Christina. The letter was not published in print. Galileo's intent was for the letter to circulate only privately among scientists, theologians, and church authorities so as to prevent the condemnation of Copernicanism by Catholic authorities. It is very likely that Galileo changed the text of the letter many times to suit specific addressees while also taking into account reactions that the letter had gradually garnered. This explains why we possess at least two quite distinct recensions of the letter to Christina, a longer one and a shorter one. Between February and May 1615 a great number of written and oral exchanges took place between Galileo, his friends, and his adversaries (part of the dossier of these texts has now been republished in Bucciantini and Camerota, *Scienza e religione*). A genuine critical edition of the letter has not yet been produced. Citations in this chapter are not from Antonio Favaro's national edition of Galileo's works (1890–1909), but from the first Strasbourg edition (1636), which Franco Motta republished in 2000 (Galilei, *Lettera a Cristina di Lorena*).

32. Pesce, "L'indisciplinabilità del metodo," 162–65.

> I should like to ask these very prudent Fathers to agree to examine very diligently the difference between debatable and demonstrative doctrines. Keeping firmly in mind the compelling power of necessary deductions, they should come to see more clearly that it is not within the power of the practitioners of demonstrative sciences to change opinion at will, choosing now this and now that one; that there is a great difference between giving orders to a mathematician or a philosopher and giving them to a merchant or a lawyer; and that demonstrated conclusions about natural and celestial phenomena cannot be changed with the same ease as opinions about what is or is not legitimate in a contract, in a rental, or in commerce.[33]

Galileo does not, however, confer an all-encompassing domain to science. He distinguishes two separate areas: that of nature, subject to scientific research, and that of faith and morality, reserved to theology, with its two main pillars: Scripture and Church Tradition.

The letter to Christina also proposed a mechanism for resolving instances of conflict between science and theology. When science demonstrates with absolute certainty a proposition that seems to be contradicted by the literal words of the Scripture,[34] theologians ought to change their interpretation of Scripture while scientists ought not to be required to revise their scientific conclusion. The unity of the entire cultural system remains guaranteed by the fact that the two domains of truth, nature and Scripture, both come from the only Divine Word. Galileo's proposed understanding of the scientific method allows a particular domain—nature—to be known independent of ecclesial power or consultation.

> I would say that the authority of Holy Scripture aims chiefly at persuading men about those articles and propositions which, surpassing all human reason, could

33. Galileo to Christina, trans. Finocchiaro, *The Galileo Affair*, 101.

34. Galileo specifies the limits of the validity of his statement: it concerns just those biblical statements that are related to natural questions, taken in the literal sense and unanimously interpreted by the ecclesiastical tradition.

> not be discovered by scientific research or by any other means than through the mouth of the Holy Spirit himself. Moreover, even in regard to those propositions which are not articles of faith, the authority of the same Holy Writ should have priority over the authority of any human writings containing pure narration or even probable reasons, but no demonstrative proofs; this principle should be considered appropriate and necessary inasmuch as divine wisdom surpasses all human judgment and speculation. However, I do not think one has to believe that the same God who has given us senses, language, and intellect would want to set aside the use of these and give us by other means the information we can acquire with them, so that we would deny our senses and reason.[35]

The extraordinary importance of Galileo's response to his adversaries' accusations is that he had the courage and the intellectual capacity to reexamine thoroughly the cultural physiognomy of what had been a fundamental base of Christian culture, the Bible. Galileo does not deny the truth or the importance of the Bible (such a denial would have been heretical). Instead he confines its authority to one limited area: faith and morals.[36] In the scientific domain, however, the Bible has no value and theology no longer possesses a universal competence. Galileo thus sought to effectively limit the cultural certainty of Scripture in two ways: in its object (being confined to faith, morals, and salvation) and in the

35. Galileo to Christina, trans. Finocchiaro, *The Galileo Affair*, 93–94.

36. Galileo did not, therefore, address at all the question of the correct theological interpretation of the Bible. The idea that theology must respect the epistemological canons of the new science is completely foreign to him. In the interpretation of the Bible, theology is completely autonomous. Galileo is thus very far from Thomas Hobbes's attempt to provide a correct interpretation of the Bible. As is well known, Hobbes intended not only to establish the real authors of the biblical books, but also to establish the meaning of biblical concepts such as Spirit, Kingdom of God, and Church. Also far from Galileo was the attitude of the Cartesian philosophers, such as Ludwig Meyer, who proposed in 1666 a philosophical hermeneutics for the correct interpretation of the Bible. See Preus, *Spinoza*, 34–66.

form of knowledge (because all that can be scientifically proven is no longer subject to the Bible's authority).

The reason for the limited validity of the Bible lies in the fact that, unlike scientific knowledge about nature, it depends on an uncertain epistemological base: "not every scriptural assertion is bound to obligations as severe as every natural phenomenon."[37] In substance, there is an epistemological dishomogeneity between Scripture and nature.

Therefore, the whole theory of the letter to Christina gives the Bible a radically new cultural profile. The Bible no longer has the function of providing cultural certainties about the knowledge of nature. However, it continues to be the bearer of absolute truth—but only in the field of faith. It provides cultural certainties only in the field of religion. Religion itself has certain limits: it no longer has cognitive competence in the field of nature. From the epistemological viewpoint, the Bible is a set of statements that are uncertain and deprived of cognitive value. Its authority is limited to knowledge that cannot be reached through reason, the senses, and scientific instruments of investigation.

A second source of cultural certainty that completely ignores the Bible and theology exists now in modern society. The Bible is no longer the same as before: it does not embrace the whole of society, but only a core of regional knowledge.

Galileo's proposal also implied a second way in which the physiognomy of the Bible was altered. By being excluded from the domain of scientific inquiry as a source of scientific knowledge about the natural world, the Bible could gain a renewed autonomy from the cultural systems in which it had become so deeply

37. Galileo to Christina, trans. Finocchiaro, *The Galileo Affair*, 93–94. In the surrounding context Galileo writes: "In disputes about natural phenomena one must begin not with the authority of scriptural passages but with sensory experience and necessary demonstrations" (ibid., 93). In his letter to Castelli he writes: "Whatever sensory experience places before our eyes or necessary demonstrations prove to us concerning natural effects should not in any way be called into question on account of scriptural passages whose words appear to have a different meaning, since not every statement of the Scripture is bound to obligations as severely as each effect of nature" (Galileo to Castelli, ibid., 50).

enmeshed. It became possible to recognize more clearly that the Bible is not Aristotelian[38] and that its conceptions differ from those of ancient Greek philosophy.[39] Thus a new possibility was opened up to distinguish between the Bible and the theological systems that are an interpretation of it. Indeed, a historical-religious understanding of the Bible itself became possible.

III. The Letter to Christina as Answer to the Letter of Cardinal Bellarmine to Foscarini

On April 12, 1615, almost a year before heliocentrism was declared heretical by a committee of theological consultants of the Congregation of the Holy Office on February 26, 1616,[40] Cardinal Bellarmine wrote the famous letter to Paolo Antonio Foscarini, which was implicitly addressed also to Galileo.[41] The aim of the

38. Galileo had already affirmed in 1612 the non-Aristotelian character of biblical affirmations about nature in two unpublished forewords to *Sidereus Nuncius* (Pesce, *L'ermeneutica biblica di Galileo*, 24–27).

39. See, e.g., part 4 of Thomas Hobbes's *Leviathan*, "Of the Kingdom of Darkness," which several decades later made a radical distinction between biblical and Greek philosophical conceptions.

40. "Censura facta in S.to Officio Urbis, die Mercurii 24 Febrarii 1616, coram infrascriptis Patribus Theologis. Prima: Sol est centrum mundi, et omnino immobilis motu locali. Censura: Omnes dixerunt dictam propositionem esse stultam et absurdam in philosophia et formaliter haereticam, quatenus contradicit expresse sententiis sacrae Scripturae in multis locis secundum proprietatem verborum et secundum communem expositionem et sensum Sanctorum Patruum et theologorum doctorum. 2a: Terra non est centrum mundi nec immobilis, sed secundum se totam movetur, etiam motu diurno. Censura: Omnes dixerunt, hanc propositionem recipere eandem censuram in philosophia; et spectando veritatem theologicam, ad minus esse in fide erroneam" (Pagano and Luciani, *I documenti del processo di Galileo Galilei*, 99–100; see also Pesce, *L'ermeneutica biblica di Galileo*, 121–2). The declaration of the theologians was not followed by a condemnation by the Holy Office, but only by a decree of the Congregation of the Index on March 5, 1616: "censuit dictos Nicolaum Copernicum *De revolutionibus orbium coelestium* et Didacum a Stunica in Job, suspendendos esse donec corrigantur. Librum vero Patris Pauli Antonii Foscarini Carmelitæ omnino prohibendum atque damnandum" (quoted in Semeria, "Storia di un conflitto," 391).

41. For an English translation, see Bellarmine to Foscarini, trans.

letter was to give them suggestions on how to avoid a clash between scientific research and theology and a condemnation of heliocentrism. He suggested that they follow the same tactic used by Andreas Osiander in his foreword to Copernicus's posthumously published *De revolutionibus* (Nuremberg, 1543). Osiander argued that Copernicus wanted to present heliocentrism only as a hypothesis, without asserting its ontological truth. In this way the new astronomical theories could be more easily accepted by Protestants who affirmed the absolute inerrancy of the Bible.

In recent decades some scholars have tried to argue that Bellarmine proposed regarding heliocentrism as a hypothesis because he was convinced that experimental scientific knowledge about nature could not arrive at ontological truth but only at a cognitive hypothesis.[42] Consequently, Bellarmine's letter to Foscarini has often been studied from the point of view of the epistemology of science. However, this methodological aspect, while certainly present, appears to be secondary. Bellarmine's proposal to consider heliocentrism as a mere hypothesis is not his only assertion in the letter and indeed it is not even the most important one. The cardinal lists at least six other issues: the truth of faith, the freedom of scientific research, the scientific truth of the Bible, the agreement between scientific opinions and faith, the danger of scientific opinions for the faith, and the dominant peripatetic philosophy and theology.

The question we must ask is why considering heliocentrism a hypothesis was important to Bellarmine for the defense of the Roman Catholic political-religious system of his time. My answer is that hypotheticism helped to reconcile the freedom of scientific

---

Finocchiaro, *The Galileo Affair*, 67–69. For the Italian, see Bellarmino a Foscarini, in Galilei, *Opere*, ed. Favaro, 12:171–2.

42. John Paul II, "Discorso alla Pontificia Academia delle Scienze," 23–34; Carroll, "Galileo, Science and the Bible," 25–7; Langford, *Galileo, Science, and the Church*, 78; Martínez, "Il significato epistemologico del caso Galileo," 45–74; Basti, *Filosofia della natura e della scienza - I*, 9–21; and Pera, "The God of Theologians and the God of Astronomers," 367–87. For an history of the question, see Damanti, *Libertas philosophandi*, 374–87; and Motta, "Epistemologie cardinalizie," 95–103.

research with the truth of faith and the Bible. It was a way to avoid "dangers" to a system based on the agreement between peripatetic philosophy and faith. Bellarmine's letter, therefore, should not be seen primarily as an expression of an epistemological theory but as a pastoral warning, the advice of a spiritual father to his spiritual son (understood to be Foscarini, Galileo, or both). It should be seen more as a document bearing on ecclesiastical history than as one bearing on the history of epistemology.

In fact, the vocabulary that Bellarmine uses in the letter, as I have tried to show elsewhere, is political and religious, and not primarily epistemological. The letter widely employs a language whose concern primarily is the practice of life, not theory. It uses expressions such as "danger," "a very dangerous thing," "proceeding prudently," "to irritate," "sense of prudence," "tolerate," "to harm," "that is sufficient," and "one would have to proceed with great care."[43] Let us focus on the word "danger." Bellarmine speaks from the point of view of the "danger for the faith."[44] What primarily worries him is the danger "to harm the holy faith by rendering Holy Scripture false."[45] He writes:

> For there is no danger in saying that, by assuming the earth moves and the sun stands still, one saves all the appearances better than by postulating eccentrics and epicycles; and that is sufficient for the mathematician. However, it is different to want to affirm that in reality the sun is at the center of the world and only turns on itself without moving from east to west, and the earth is in the third heaven and revolves with great speed around the sun; this is a very dangerous thing.[46]

The first danger for Bellarmine is political; it is the danger "to irritate all scholastic philosophers and theologians."[47] The second

43. Bellarmine to Foscarini, trans. Finocchiaro, *The Galileo Affair*, 67–69.

44. Pesce, "L'indisciplinabilità del metodo," 151–74.

45. Bellarmine to Foscarini, trans. Finocchiaro, *The Galileo Affair*, 67.

46. Ibid.

47. Bellarmine to Foscarini, trans. Finocchiaro, *The Galileo Affair*, 67. Bucciantini and Camerota also agree on this point (*Scienza e religione*, xxx–xxxi and 157).

danger is related to faith; it is the danger "to harm the holy faith by rendering the Holy Scripture false."[48] The meaning of the term "danger" in Bellarmine is clear also in the light of other uses of the term in his works.[49] The cardinal uses the word when political events create an indirect danger to the faith or when Protestant theories create danger for the Catholic Church. Bellarmine speaks of the battle against the "heresy" undertaken *non sine magno . . . periculo*. He speaks also of *periculosissima tempora* in relation to the spread of Protestantism. A law can indirectly be a danger in a religious matter: *res animarum periculum concernens*. A heretic king constitutes a danger: *tolerare regem haereticum . . . est exponere religionem evidentissimo periculo*.

Bellarmine's first point in the letter to Foscarini suggests prudence: one must not "irritate" philosophers and theologians, otherwise they become dangerous. He warns Galileo about the repressive reactions that could be implemented against him by ecclesiastic Catholic authorities. In a second point, Bellarmine writes: "Consider now, with your sense of prudence, whether the Church can tolerate giving Scripture a meaning contrary to the Holy Fathers and to all the Greek and Latin commentators."[50] The whole problem lies, then, in finding a solution that the church can "tolerate." This is what Bellarmine calls "prudence."[51]

In the third point of the letter, Bellarmine expresses theological theories, but his concern is mainly practical. He recommends a way to present the agreement between indubitable scientific truth and Scripture: "One would have to proceed with great care in explaining the Scriptures that appear contrary; and say rather that we

48. Bellarmine to Foscarini, trans. Finocchiaro, *The Galileo Affair*, 67.

49. I would like to thank Franco Motta for the indication of the use of the term "danger" in Bellarmine, outside his letter to Foscarini (personal communication).

50. Bellarmine to Foscarini, trans. Finocchiaro, *The Galileo Affair*, 68.

51. Federico Cesi also spoke of a kind of dissimulation ("temperamento"): "Conobbi seco che ragionevolmente i revisori dovevano restare soddisfatti del temperamento che V.S. mi mandò" (Federico Cesi to Galileo, November 30, 1612, in Galilei, *Opere*, ed. Favaro, 5:439; see also Pesce, *L'ermeneutica biblica di Galileo*, 24–27).

do not understand them than that what is demonstrated is false."[52] In other words, Bellarmine's practical suggestion to Galileo is to avoid affirmations that could be theologically dangerous. Bellarmine teaches Galileo how he should write in order to avoid being condemned. It is as if Bellarmine comes out of his public functions to give private advice to Foscarini and Galileo. He advises them to assume an attitude that can preserve opposing needs while not disturbing the system of power in Catholic society.

Bellarmine wrote his letter from the point of view of the political-religious system of Roman Catholicism to which he belonged. His concern was to defend the Bible as a cultural certainty and as the foundation of a traditional symbolic universe in which all Christian theology, as well as the power of the church over society, had their place. His aim was to maintain theology's authority to extend to all aspects of social reality.

For these reason it seems to me that Bellarmine's proposal to consider heliocentrism a hypothesis cannot be seen as an anticipation of contemporary epistemological theories. Its main aim is to preserve the overall cultural function of the Bible. Bellarmine reaffirms the cultural certainty and the universal truth of the Bible in every field. For example, he asserts that the Bible must be accepted as historically true in all its details. One must not doubt, for example, that Abraham had two sons:

> Nor can one answer that this is not a matter of faith, since if it is not a matter of faith as regards the topic, it is a matter of faith as regards the speaker; and so it would be heretical to say that Abraham did not have two children and Jacob twelve, as well as to say that Christ was not born of a virgin, because both are said by the Holy Spirit through the mouth of the prophets and the apostles.[53]

Similarly, Bellarmine affirmed the scientific reliability of the Bible because Solomon received revelations in the field of natural knowledge. Bellarmine's main concern is thus to maintain the

52. Bellarmine to Foscarini, trans. Finocchiaro, *The Galileo Affair*, 68.

53. Ibid. Regarding Bellarmine's understanding of the Bible's verbal inspiration, see Motta, *Bellarmino*, 531–32.

truth of the Bible in every area of life and knowledge by affirming that it is a universal source of cultural certainty. Bellarmine certainly admits, in principle, that the Bible could express itself in a way that would contradict scientific statements that had actually been proven to be true. In this way he implicitly acknowledges that science can make ontologically true statements, and therefore he admits implicitly the existence of an alternative source of certainty that is independent of the Bible. Nevertheless, he will not admit that the certainty of the Bible should be restricted to a particular sector of knowledge. To be sure, he does leave open a very small window of opportunity, but he is quick to say that he considers this eventuality to be quite impossible. We are at the antipodes of Galileo.

Galileo held Bellarmine's observations in high regard and his letter to Christina constitutes an indirect response to him.[54] He was, however, at the antipodes of Bellarmine's way of thinking about the relationship between science and theology in the society of their time.

IV. The Epistemological Dishomogeneity of Bible and Science

To be sure, it would be hard to affirm that prior to the seventeenth century no one searched for an alternative source of certainty with respect to the cultural certainties of a society. Greek philosophy from its very beginning presents itself as a critical (or at least dialectical) instrument in relation to traditional religious knowledge. In her book *La Raison de Rome*, Claudia Moatti has written about the birth in late Republican Rome of a complex movement of thought that tried to call into question religious traditions and traditional ways of thinking.[55] And even medieval philosophy never ceased looking for an alternative source of certainty that would

54. Regarding the thesis that the letter to Christina is a response to Bellarmine, see Pesce, *L'ermeneutica biblica di Galileo*, 97–98.

55. Moatti, *La raison de Rome*, 173–88.

differ from traditional certainties available in different societies of the time.

What happens, however, with the Galilean experimental method is qualitatively different. With Galileo, the totality of traditional knowledge is deprived of its epistemological certainty. Furthermore, in the letter to Christina, it is science and not philosophy that drives the epistemological revolution. A confirmation of this lies in the fact that Galileo does not resort to the philosophical principles of Paduan Aristotelians. He refuses an approach similar to the one taken, for example, by Pietro Pomponazzi in his criticism of religions. Indeed, Galileo refutes the position of the double truth by affirming his commitment to the principle established by the Fifth Lateran Council against this theory. Galileo also professes his adherence to the Council of Trent's principle of interpretation which reserved to the Church the interpretation of the Bible. For Galileo, the Bible remains a source of certainty but only in a limited area and without basic epistemological certainty. The cultural function of the Bible has changed. From this moment, I repeat, the Bible is not the same anymore.

F. David Peat's book, *From Certainty to Uncertainty: The Story of Science and Ideas in the Twentieth Century*, shows that scientists themselves have criticized the belief that science can arrive at certainty. But awareness of the epistemological limits of contemporary science does not lead at all to the rehabilitation of prescientific astronomy, physiology, or Aristotelian physics. The ancient cultural systems with which the different books of the Bible were inextricably mixed remain entirely useless from a scientific point of view. Conversely, what still remains completely intact is the basic prerequisite of the scientific method, namely, that it cannot base itself on any traditional cultural certainty. Peat writes in his book that it is legitimate to ask how much of western science "is inevitable and objective, and how much is culturally conditioned and determined."[56] This same author, however, writes that "Science asserts that the answers nature provides are independent of

56. Peat, *From Certainty to Uncertainty*, 209.

culture, belief, and personal values. Cultural relativism - it argues - has no place in science."[57]

We can therefore affirm what George Coyne and Michael Heller wrote many years ago:

> [The] use of the Bible to derive scientific knowledge [had] deplorable consequences . . . However, in Augustine's time such a practice was unavoidable. In modern times the natural sciences (geology, biology, cosmology, etc.) stimulate the purification of biblical exegesis from the literal understanding of the world image presented by biblical texts.[58]

Obviously this answer leaves unanswered an essential point: the relation of the results of the scientific research to the social context and to the natural environment. This is not a question of science, but of ethics and politics. In conclusion, from my point of view, the most important question at stake was not whether Galileo had really proven the truth of the Copernicanism, but whether or not the Bible could be considered the principal foundation of cultural certainty in modern societies.

Bibliography

Artigas, Mariano, and Melchor Sánchez de Toca. *Galileo e il Vaticano: Storia della Pontificia Commissione di Studio sul Caso Galileo (1981–1992)*. Translated by Maria Pertile. Venice: Marcianum, 2009.

Basti, Gianfranco. *Filosofia della natura e della scienza—I*. Rome: Lateran University Press, 2002.

Beretta, Francesco. *Galilée devant le tribunal de l'inquisition: Une relecture des sources*. Fribourg, Switzerland: Université de Fribourg, 1998.

———. "Le procès de Galilée et les Archives du Saint-Office: Aspectes judiciaires et théologiques d'une condamnation célèbre." *Revue des Sciences Philosophiques et Théologiques* 83 (1999) 441–90.

———. "'Omnibus Christianae, Catholicaeque Philosophiae amantibus. D. D.': Le *Tractatus syllepticus* de Melchior Inchofer, censeur de Galilée." *Freiburger Zeitschrift für Philosophie und Theologie* 48 (2001) 301–27.

57. Ibid., 208.

58. Coyne and Heller, *A Comprehensible Universe*, 51.

Brandmüller, Walter. *Galilei e la chiesa: Ossia il diritto ad errare*. Scienza e fede 4. Vatican City: Libreria Editrice Vaticana, 1992.

Bucciantini, Massimo, and Michele Camerota, eds. *Scienza e religione: Scritti copernicani*. Rome: Donzelli, 2009.

Campanella, Tommaso. *Apologia pro Galileo, mathematico florentino. . . .* Frankfurt, *typis Erasmi Kempfferi,* 1622.

Carroll, William E. "Galileo, Science and the Bible." *Acta Philosophica* 6 (1997) 5–37.

Coyne, George V. "The Church's Most Recent Attempt to Dispel the Galileo Myth." In *The Church and Galileo*, edited by Ernan Mcmullin, 340–59. Notre Dame: University of Notre Dame Press, 2005.

Coyne, George V., and Michael Heller. *A Comprehensible Universe: The Interplay of Science and Theology*. New York: Springer, 2008.

Damanti, Alfredo. *Libertas philosophandi: Teologia e filosofia nella lettera alla granduchessa Cristina di Lorena di Galileo Galilei*. Temi e testi 71. Rome: Edizioni di storia e letteratura, 2010.

Descartes, René. *Meditations on First Philosophy: With Selections from the Objections and Replies; A Latin-English Edition*. Translated by John Cottingham. Cambridge: Cambridge University Press, 2013.

Fantoli, Annibale. "Il caso Galileo: Una questione chiusa?" In *Il processo a Galileo Galilei e la questione galileiana*, edited by Gian Mario Bravo and Vincenzo Ferrone, 229–57. Rome: Edizioni di Storia e Letteratura, 2010.

———. *Galileo: Per il copernicanesimo e per la chiesa*. 3rd ed. Vatican City: Libreria Editrice Vaticana, 2010.

———. "Galileo e il Vaticano: Una nuova storia documentaria dei lavori della Commissione galileiana e dei discorsi di chiusura (1981–1992)." *Galileiana* 6 (2009) 273–97.

Ferrone, Vincenzo. *Scienza natura religione: Mondo newtoniano e cultura italiana nel primo settecento*. Storia e diritto 9. Naples: Jovene, 1982.

Finocchiaro, Maurice A., ed. and trans. *The Galileo Affair: A Documentary History*. Berkeley: University of California Press, 1989.

———. *The Trial of Galileo: Essential Documents*. Indianapolis: Hackett, 2014.

Fomiano, Salvatore, ed. and trans. *Apologia per Galileo*, by Tommaso Campanella. Milan: Marzorati, 1971.

Galilei, Galileo. *Dialogue Concerning the Two Chief World Systems*. Translated by Stillman Drake. Berkeley: University of California Press, 1953.

———. *Lettera a Cristina di Lorena: Sull'uso della Bibbia nelle argomentazioni scientifiche*. Edited by Franco Motta. Genoa: Marietti, 2000.

———. *Le opere di Galilei: Edizione nazionale*. Edited by Antonio Favaro. 20 vols. Florence: Giunti Barbèra, 1890–1909. Reprinted, 1968.

Galluzzi, Paolo. "Il 'caso Galileo.'" In *Il processo a Galileo Galilei e la questione galileiana*, edited by Gian Mario Bravo and Vincenzo Ferrone, 3–16. Rome: Edizioni di Storia e Letteratura, 2010.

Gassendi, Pierre. *Apologia in Io. Bap. Morini librum, cui titulus, Alae Telluris Fractae. . . .* Lyon: Barbier, 1649.

Giannone, Pietro. "Parere intorno la censura del padre Massimiliano Galler gesuita sopra il libro di Giovan Paolo Ganzer dottore in medicina e filosofia." In vol. 1 of *Pietro Giannone e il suo tempo: Atti del convegno di studi nel tricentenario della nascita; Foggia-Ischitella, 23–24 ottobre 1976*, edited by Raffaele Ajello, 367–413. 2 vols. Naples: Jovene, 1980.

Gregory, Tullio, "Filosofia e teologia nella crisi del XIII secolo." In *Mundana sapientia: Forme di conoscenza nella cultura medievale*, 61–76. Rome: Edizioni di Storia e Letteratura, 1992.

Guthke, Karl S. *The Last Frontier: Imagining Other Worlds, from the Copernican Revolution to Modern Science Fiction*. Ithaca, NY: Cornell University Press, 1990.

Inchofer, Melchior. *Tractatus syllepticus, in quo, quid de terrae, solisque, motu, vel statione secundum S. Scripturam. . . . breviter ostenditur.* Rome: Ludovicus Grignanus,1633.

John Paul II. "Discorso alla Pontificia Academia delle Scienze, 31 ottobre 1992." In *La nuova immagine del mondo: Il dialogo tra scienza e fede dopo Galileo*, edited by Paul Poupard, 23–34. Casale Monferrato: Piemme, 1996.

Langford, Jerome J. *Galileo, Science, and the Church*. Rev. ed. Ann Arbor: University of Michigan Press, 1966.

Martínez, Rafael. "Il significato epistemologico del caso Galileo: Due diverse concezioni della scienza." *Acta Philosophica* 3 (1994) 45–74.

Moatti, Claudia. *La raison de Rome: Naissance de l'esprit critique à la fin de la République (IIe-Ier siècle avant Jésus-Christ)*. Paris: Seuil, 1997.

Motta, Franco. *Bellarmino: Una teologia politica della Controriforma*. Brescia: Morcelliana, 2005.

———. "Epistemologie cardinalizie: Ipotesi, verità, apologia." In *Il processo a Galileo Galilei e la questione galileiana*, edited by Gian Mario Bravo and Vincenzo Ferrone, 95–103. Rome: Edizioni di Storia e Letteratura, 2010.

Pagano, Sergio M., ed. *I documenti vaticani del processo di Galileo Galilei (1611–1741)*. Collectanea Archivi Vaticani 69. Vatican City: Vatican Secret Archives, 2009.

Pagano, Sergio M. and Antonio G. Luciani, eds. *I documenti del processo di Galileo Galilei*. Collectanea Archivi Vaticani 21 and Scripta Varia 53. Vatican City: Pontificia Academia Scientiarum, 1984.

Peat, F. David. *From Certainty to Uncertainty: The Story of Science and Ideas in the Twentieth Century*. Washington, DC: Joseph Henry, 2002.

Pera, Marcello. "The God of Theologians and the God of Astronomers: An Apology of Bellarmine." In *The Cambridge Companion to Galileo*, edited by Peter Machamer, 367–87. Cambridge Companions to Philosophy. Cambridge: Cambridge University Press, 1998.

Pesce, Mauro. "Il Consensus Veritatis di Christoph Wittich e la distinzione tra verità scientifica e verità biblica." *Annali di storia dell'esegesi* 9 (1992) 53–76.

———. "Il Copernicanesimo e la teologia: Perché il 'caso' Galileo non è chiuso." In *Il caso Galileo: Una rilettura storica, filosofica, teologica; convegno internazionale di studi, Firenze, 26–30 maggio 2009*, edited by Massimo Bucciantini, Michele Camerota, and Franco Giudice, 33–46. Biblioteca di Galilæana 2. Florence: Olschki, 2011.

———. *L'ermeneutica biblica di Galileo e le due strade della teologia cristiana*. Uomini e dottrine 43. Rome: Edizioni di Storia e Letteratura, 2005.

———. "Galileo a Cristina," in *Scienza e religione: Scritti copernicani*, edited by Massimo Bucciantini and Michele Camerota, 35–84. Rome: Donzelli, 2009.

———. "L'indisciplinabilità del metodo e la necessità politica della simulazione e della dissimulazione in Galilei dal 1609 al 1642." In *Disciplina dell'anima, disciplina del corpo e disciplina della società tra medioevo ed età moderna: Convegno Internazionale di Studio*, Bologna 7–9 ottobre 1993, edited by Paolo Prodi and Carla Penuti, 151–74. Annali dell'Istituto Storico Italo-Germanico 40. Bologna: Mulino, 1994.

———. "Gli ingegni senza limiti e il pericolo per la fede." In *Largo campo di filosofare: Eurosymposium Galileo 2001*, edited by José Montesinos and Carlos Solís, 637–59. La Orotava, Spain: Fundación Canaria Orotava de Historia de la Ciencia, 2001.

———. "L'interpretazione della Bibbia nella lettera di Galileo a Cristina di Lorena e la sua ricezione: Storia di una difficoltà nel distinguere ciò che è religioso da ciò che non lo è." *Annali di storia dell'esegesi* 4 (1987) 239–84.

———. "La 'Lettera a Cristina': Una proposta per definire ambiti autonomi di sapere e nuovi assetti di potere intellettuale nei paesi cattolici." Introduction to *Lettera a Cristina di Lorena: Sull'uso della Bibbia nelle argomentazioni scientifiche*, by Galileo Galilei, 7–66. Edited by Franco Motta. Genoa: Marietti, 2000.

———. "Momenti della ricezione dell'ermeneutica biblica galileiana e della Lettera a Cristina nel XVII secolo." *Annali di storia dell'esegesi* 8 (1991) 55–103.

———. "Una nuova versione della lettera di G. Galilei a B. Castelli." *Nouvelles de la République des lettres* 11 (1991) 89–122.

———. "Le redazioni originali della lettera 'Copernicana' di G. Galilei a B. Castelli." *Filologia e critica* 17 (1992) 394–417.

Poupard, Paul. "La controversia tolemaico-copernicana nei secoli XVI e XVII." In *La nuova immagine del mondo: Il dialogo tra scienza e fede dopo Galileo*, edited by Paul Poupard, 15–21. Casale Monferrato, Italy: Piemme, 1996.

Preus, J. Samuel. *Spinoza and the Irrelevance of Biblical Authority*. Cambridge: Cambridge University Press, 2001.

Prosperi, Adriano. "L'Inquisizione e Galilei." In *Il processo a Galileo Galilei e la questione galileiana*, edited by Gian Mario Bravo and Vincenzo Ferrone, 17–38. Rome: Edizioni di Storia e Letteratura, 2010.

Semeria, Giovanni. "Storia di un conflitto tra la scienza e la fede." *Rivista di studi religiosi* 3 (1903) 388–416.

Van Helden, Albert. *The Invention of the Telescope*. Transactions of the American Philosophical Society 67.4. Philadelphia: American Philosophical Society, 1977.

Westman, Robert S. "The Copernicans and the Churches." In *God and Nature: Historical Essays on the Encounter between Christianity and Science*, edited by David C. Lindberg and Ronald L. Numbers, 76–113. Berkeley: University of California Press, 1986.

4

Galileo's Telescope

Dennis D. McCarthy

Introduction

When Galileo Galilei turned his newly crafted *perspicillum* (later renamed telescope) toward the Moon on the evening of December 1, 1609, he started on a path leading to our current understanding of our universe. Progress on that path has provided profound changes not only in astronomy but also in many other areas of human endeavor. In order to comprehend the significance of Galileo's telescope it is first important to understand the nature of the search for knowledge in the Renaissance era of the early seventeenth century. The construction of the instrument itself reveals a bit of Galileo's spirit, and his methodical and analytical observations with his telescope disclose his curiosity about the natural world. Even during his lifetime Galileo's telescope was instrumental in developing a new approach to thinking about the natural world, the scientific method.

Background

Galileo Galilei was born on February 15, 1564 (in the Julian calendar), near Pisa,[1] and in that era in Europe the Aristotelian

1. Drake, *Galileo at Work*, 1.

tradition was the generally accepted explanation of the universe. Aristotle had distinguished between theoretical, practical, and productive sciences in his *Metaphysics*.[2] Theoretical sciences dealt with knowledge, practical sciences with conduct, and productive sciences with making useful objects. Theoretical sciences were (1) metaphysics, which was concerned with unchangeable things separate from matter or body, e.g., spiritual substances, (2) mathematics dealing with unchangeable things that have no separate existence, e.g., numbers, and (3) physics, or natural philosophy as it was normally referred to, which deals with changeable things having a separate existence.[3] Knowledge of the natural world around the dawn of the seventeenth century, then, mainly involved two disciplines: natural philosophy and mathematics. The intention of natural philosophy (*philosophia naturalis* or *scientia naturalis*) was to explain the characteristics of the natural world.[4] It involved abstract philosophizing aimed at finding truth. Peter Dear writes, "this was the rational counterpart of belief, and spoke to intellectual conviction rather than practical know-how."[5]

On the other hand, mathematics was concerned with establishing mathematical relationships corresponding to observations without any concern for physical explanations. Medieval mathematics had been largely hypothetical and unconcerned with empirical investigation.[6] By the seventeenth century the two disciplines were being united in the study of physics. Astronomy was considered as being situated between natural philosophy and mathematics. However, it continued to be more mathematical in nature being concerned mostly with the determination and prediction of the directions of stars and planets.[7] Astronomy was considered to be important for regulating the calendar, particularly for religious feasts, for timekeeping, for navigation at sea, for proper

2. Grant, *The Foundations of Modern Science*, 135.

3. Ibid.

4. Dear, *Revolutionizing the Sciences*, 3.

5. Ibid.

6. Grant, *A History of Natural Philosophy*, 311.

7. Ibid., 312.

orientation of churches, and for astrology. Also, because it could possibly reveal hidden causes in nature, astronomy could contribute to revealing God's ways. Although discouraged by the Church, astrology had become an important motivation for the study of astronomy following the plague of the fourteenth century,[8] and it did serve to be a reason to fund astronomical work. Galileo's manuscripts, for example, do contain a number of horoscopes relating to his family and friends.[9]

The natural philosopher's view of the world at that time was that the universe was a finite Earth-centered sphere populated with a number of sub-spheres associated with planets. The volume below the sphere of the Moon, the sub-lunary sphere, was considered to be corruptible, mutable and composed of the four elements: earth, air, fire, and water. A finite universe implies the existence of a center, and because earth was the heaviest of these elements, it was only natural that the Earth should be at the center, as the heaviest element would naturally be expected to fall to the center. Beyond the sphere of the Moon lay the celestial spheres composed of a single element, aether. The components of this volume were thought to be incorruptible, immutable and the planets were supposed to move on their respective spheres in perfect circular motion.[10]

For astronomer/mathematicians the heliocentric model of the solar system proposed by Nicolaus Copernicus (1473–1543) published over 70 years before in 1543 was challenging the geocentric conception of the solar system of Claudius Ptolemy (ca. 100—ca. 175) as a means to describe the motion of the Sun, Moon, and planets. The latter system is described in his *Mathematical Syntaxis* better known as the *Almagest*, which is the Latinized version of the Arabic name it received when it was translated by Islamic scholars.[11] It served astronomers well in their craft of describing and predicting the motions of the planets. The key features of the

8. Pederson, "Astronomy," 304–5.

9. Drake, *Galileo at Work*, 55.

10. Grant, *The Foundations of Modern Science*, 134.

11. Grant, *Science and Religion*, 78.

Ptolemaic solar system model were geocentrism, circular motion and the use of geometric figures that could be used to model the motion of the planets. Different geometric models could be employed in order to save the appearances of the planetary motions. These include an eccentric circle, an epicycle-on-deferent model, an equant model and finally a combination epicycle-on-deferent/equant model.[12]

The eccentric circle required a planet to execute uniform circular motion about a center that was at some distance from the Earth. The principle of epicyclic motion was that a planet follows a perfect circular path with uniform motion on an epicycle, the center of which follows perfect uniform circular motion on another circle called a deferent. To describe planetary motion in finer detail multiple epicycles were often employed. The equant model makes use of a point about which a planet travels through equal angles in equal time. The combination of the latter two models required the center of the epicycle to move about an equant (see Figure 4.1). The use of these figures enabled the calculations of the mathematician/astronomers but they were not represented as describing the physics of the problem.

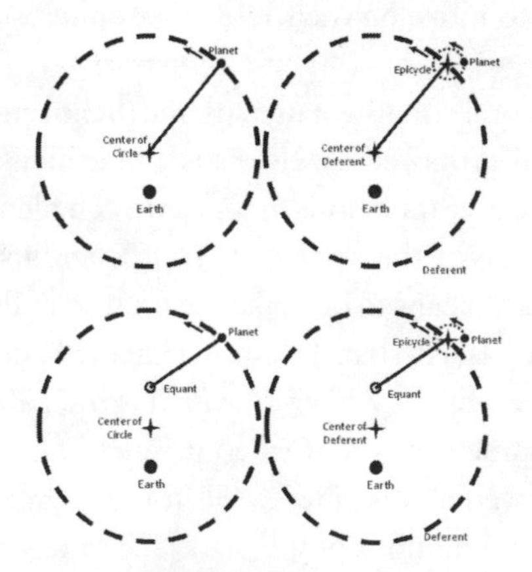

Figure 4.1. Planetary orbits in the Ptolemaic theory, top row: eccentric circle and epicycle; bottom row: equant, and combination of equant and epicycle. (Illustrations by author)

12. Ibid., 81; Lindberg, *The Beginnings of Western Science*, 99–105.

In 1543 the famous work of Copernicus *De Revolutionibus Orbium Coelestium* had been published. In that publication Copernicus resurrected the concept of the Earth moving about the Sun, that had been proposed in antiquity.[13] Placing the Sun at the center of the universe simplified the system but he was still forced to make use of some of the same geometric constructions as were used in the Ptolemaic system, but without the use of the equant.[14] The famous astronomical observer Tycho Brahe (1546–1601) had difficulty in accepting the Copernican concept of the Earth's motion for physical and religious reasons. He also was unwilling to accept the Ptolemaic system and so he proposed a compromise system in his book on the comet of 1577 *Concerning the New Phenomena in the Ethereal World* published in 1588.[15] In his system the Earth remained fixed at the center of the universe and the planets revolved about the Sun rather than the Earth. Although his proposal continued to make use of epicycles, it made use of crystalline spheres to carry the planets as had been supposed in the Ptolemaic and Copernican systems.

All of these computational models sought to describe actual observations that were made with a variety of instruments designed to measure angles in the sky. One angle typically measured was the altitude of a celestial object above the horizon. Mechanical clocks were improving to the point that they could be used to measure the longitudinal angle between objects by noting the time difference between the passage of the objects across some fixed reference marker. The most widely used astronomical instrument of the time was the astrolabe. In its most basic form it was a circular metal plate on which was inscribed a reference line. A metal pointer was attached to the plate so that it rotated about the center of the plate, and a metal ring would be attached to the top of the plate along the diameter perpendicular to the inscribed reference line. The astrolabe was used by suspending it using the attached ring. The observer would sight along the moving pointer in the

13. Dear, *Revolutionizing the Sciences*, 33–36.

14. Dreyer, *A History of Astronomy from Thales to Kepler*, 305–44.

15. Ibid., 362–65.

direction of the celestial object and read the angle between the direction of the pointer and the reference line.[16] These instruments could be quite elaborate with multiple plates being attached that rotated on the axis of the basic plate. Other instruments used to make observations included the quadrant, a metal plate cut in the shape of a quarter of a circle that could be used to measure altitudes of celestial objects. More sophisticated instruments employed were the armillary sphere and a "torquetum," an instrument designed to measure angles in the longitudinal direction as well as the angular height of an object above the plane of the Earth's equator. It was made of brass plates and sighting pointers as well. Another instrument that could be used was a "triquetrum" that could be used to measure zenith distances of celestial objects (i.e., the angular distances between the overhead point, called the zenith, and the direction to the object).[17] All of these instruments involved the observer sighting along a pointer without any optical aids.

In 1609 Galileo was 35 years old and, since 1591, had held the chair of mathematics at the University of Padua having served in that capacity before in Pisa. When he turned that *perspicillum* on the Moon in 1609 he was working within the context of the astronomical world of his day that included the conviction that the Earth was at the center of a finite spherical universe which included a region beyond the Moon where perfection reigned. Astronomer/mathematicians employed mathematical and geometrical constructions to describe the observed changing directions of celestial objects, but it was not assumed that those constructions necessarily represented physical nature. Although the Ptolemaic system was generally accepted other systems had been proposed. The heliocentric system of Copernicus was not generally accepted. Observations could not prove that the Earth was in motion about the Sun, and this concept was at odds with Aristotle's physics. It just didn't make common sense. A more serious concern was, that by suggesting that the Earth was in motion, it would imply that the Earth was actually in the heavens. That contradicted the prevailing

16. Crombie, *Medieval and Early Modern Science*, 91–96.

17. Pederson, "Astronomy," 322–31.

concepts of celestial and sublunary regions in the universe.[18] As for the theologians of the day, heliocentrism had not been a concern when it was proposed originally because it was understood to be a mathematical tool unrelated to the actual physical nature of the universe. However, the protestant reformation triggered controversy regarding the authority of the church and the nature of scriptural authority. Protestant theologians were concerned about the fact that there was no scriptural evidence for a heliocentric universe and, in fact, it seemed to be contra-indicated. Roman Catholic theologians were concerned about the erosion of their authority and balked at accepting interpretations not previously accepted by church fathers or tradition. So, in 1609 Galileo turned his *perspicillum* to the sky with all of these thoughts in mind.

The Instrument

The instrument that Galileo used that night was the result of a series of events that had begun a year before. In 1608 three patent claims were made to the States General of the province of Zeeland in the Netherlands, and because of confusion surrounding these applications, all were rejected in early December 1608. However Jacob Metius (d. 1628) and Hans Lipperhey, or Lippershey (1570–1619) were granted some funding. Metius was awarded 100 guilders to improve his instrument and Lipperhey was awarded 600 guilders to complete two more instruments.[19] News of this invention spread quickly and in the spring of 1609 spyglasses appeared in shops in Paris, and they were beginning to be found in Italy.[20] Galileo writes in his 1610 publication *Sidereus Nuncius* that Jacques Badovere (c. 1575—c. 1620), a French nobleman and a former student of Galileo had confirmed to Galileo the optical properties of the instrument.[21] He was in Venice in July, 1609, where he apparently learned of the spyglass. Galileo writes:

18. Lindberg and Numbers, *When Science & Christianity Meet*, 38–47.

19. Del Santo et al., "Galileo's Telescope," 35–37.

20. Ibid., 38.

21. Land, *The Telescope Makers*, 5; Del Santo et al., "Galileo's Telescope," 42.

> In Venice, where I happened to be at the time, news arrived that a Fleming had presented to Count Maurice [of Nassau] a glass by means of which distant objects could be seen as distinctly as if they were nearby. That was all. Upon hearing this news, I returned to Padua, where I then resided and set myself to thinking about the problem. The first night after my return I solved it, and on the following day I constructed the instrument and sent word of this to these same friends at Venice with whom I had discussed the matter the day before. Immediately afterward I applied myself to the construction of another and better one, which six days later I took to Venice, where it was seen with great admiration by nearly all the principal gentlemen of that Republic for more than a month on end, to my considerable fatigue.[22]

It is not clear whether Galileo received news of the spyglass from Badovere or from another friend, Paolo Sarpi (1552–1623), or if he might have actually possessed one from Paris or Venice.[23] In any event, by August of 1609 Galileo had constructed his own version.

The optics of Galileo's spyglass were made up of two lenses, a plano-convex lens for the objective and a plano-concave lens for the ocular (see Figure 4.2).[24] Figure 4.3 shows schematically the path of the light rays and the magnifying properties of the construction.

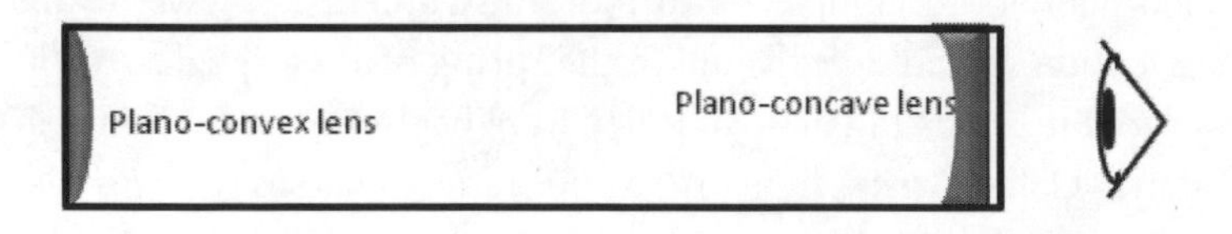

Figure 4.2. Schematic view of Galileo's spyglass. (Illustration by author).

22. Drake, *Galileo at Work*, 137–38.
23. Del Santo et al., "Galileo's Telescope," 42.
24. Land, *The Telescope Makers*, 9.

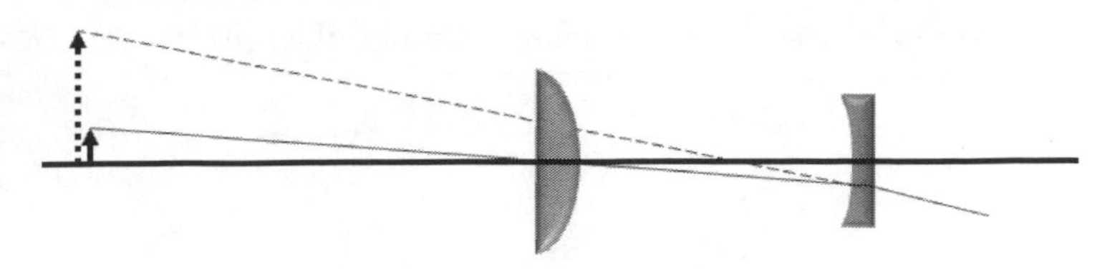

Figure 4.3. Optics of Galileo's spyglass. (Illustration by author.)

The ocular lens is placed in front of the focal point of the objective at a distance from the focal point equal to the focal length of the eyepiece. The image appears to be erect with this design, but it provides a very narrow field of view and a maximum power that can be achieved practically of about 30X. His first attempt resulted in an instrument of about 3X power, but he was able to achieve 8X power with the second, made soon after the first,[25] and this was the instrument that he took to Venice, where he donated it to the Doge, Leonardo Donato (1536–1612).[26] At that time Galileo was a professor of mathematics at the University of Padua, which was in the Venetian Republic. For this Galileo was rewarded with a lifetime appointment at that University and the promise of a significant increase in salary.[27] Apparently even though he was rewarded for his work by the Venetian senate, he re-opened negotiations with the Tuscan court in Florence where his former pupil, Cosimo de' Medici had become Grand Duke Cosimo II de' Medici.[28]

In November, 1609, he constructed a third instrument, capable of about 20X power and this was the *perspicillum* used on December 1, 1609, to look at the Moon.[29] Galileo went on to make a number of lenses[30] for which glass of good quality was required.

25. Drake, *Galileo at Work*, 139.
26. Del Santo et al., "Galileo's Telescope," 43.
27. Drake, *Galileo at Work*, 141.
28. Ibid., 142.
29. Ibid., 143.
30. Drake, *Galileo at Work*, 281.

He tried to overcome the rather poor condition of early seventeenth century glass by buying a large number of lenses.[31]

Figure 4.4. Galileo's telescopes. Two of Galileo's telescopes appear above the framed lens of the instrument presented to Cosimo II de' Medici.[32] The lens was accidentally cracked later. (Image in public domain)

The instrument did not receive the name "telescope" until 1611 when Galileo presented one to the Accademia dei Lincei in Rome on the occasion of his induction into that society. Upon the suggestion of Giovanni Demisiani (d. 1614), Prince Federico Cesi (1585–1630), the founder of that society gave it the name by which it is now commonly known. The word itself is a combination of the Greek words *tele* ("at a distance") and *skopeo* ("I observe").[33]

31. Mando et al., "The Quality of Galileo's Lenses," 67.
32. Fahre, "The Scientific Works of Galileo."
33. Del Santo et al., "Galileo's Telescope," 45; Drake, *Galileo at Work*, 196.

The Observations

Although Galileo was not the first person to look at objects in the sky with a telescope, he did begin a series of investigations into celestial objects that had never been conceived before. The observations that he made covered the obvious objects of interest such as the Moon, the Sun, the planets and the stars but the telescope permitted this work to be done with unprecedented capability. This capability, combined with Galileo's spirit and intellect, had far-reaching effects.

The Moon

Galileo's telescopic observations of the Moon began on December 1, 1609.[34] On July 26, 1609 (in the Julian calendar), however, an English scientist, Thomas Harriot (1560–1621), apparently had already made the first telescopic lunar observation, which he documented in a drawing (see Figure 4.5). He was unsure of what he actually was seeing, and the quality of his instrument was probably too poor to draw any conclusion regarding the surface of the Moon.[35] He subsequently mentioned that he had "observed a strange spottednesse" (*sic*).

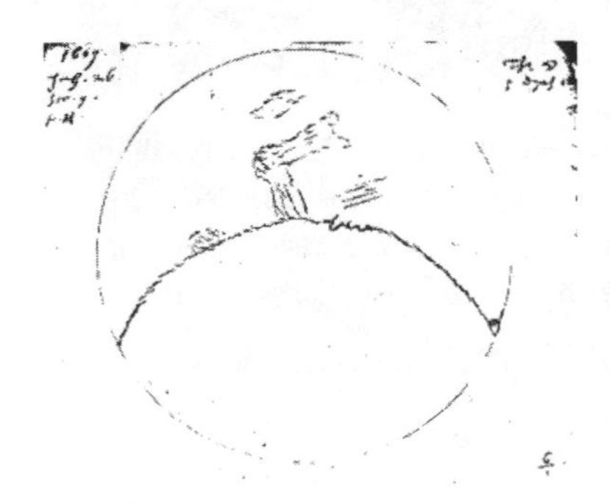

34. Drake, *Galileo at Work*, 143.

35. Bloom, "Borrowed Perceptions: Harriot's Maps of the Moon," 117–19.

Figure 4.5. Thomas Harriot's drawing of the Moon on 5 August 1609 (Gregorian calendar) (left)[36] and image of the Moon for the same date. (Graphics courtesy of Starry Night® Pro 7/ Simulation Curriculum Corp.)

Galileo published his observations in 1610 in his work *Sidereus Nuncius*. Figure 4.6 shows a series of lunar images including a black and white version of Galileo's watercolor sketches of the Moon in December, 1609,[37] an image of the Moon from *Sidereus Nuncius*,[38] and an image of the Moon as it would actually appear which was made using simulation software.

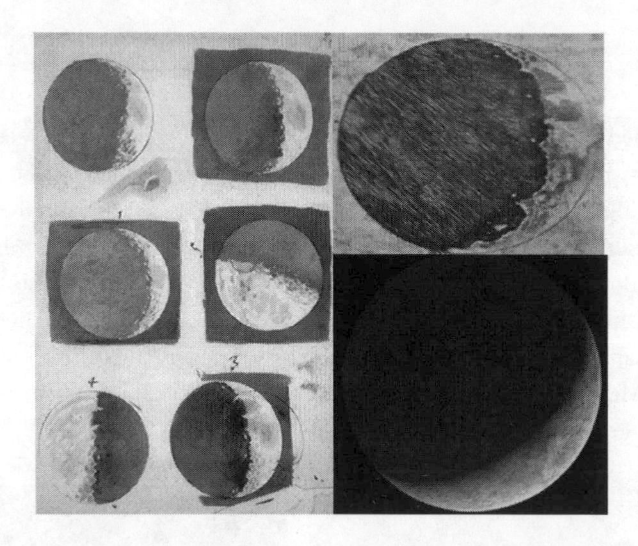

Figure 4.6. Black and white version of Galileo's water color sketches of the Moon in December 1609[39] (left), an image of the Moon from *Sidereus Nuncius*[40] (top right) and simulation of the lunar appearance on December 1, 1609. (Graphics courtesy of Starry Night® Pro 7/ Simulation Curriculum Corp.)

36. Ibid., 118.
37. "Open Culture."
38. Galilei, 1610. "Sidereus Nuncius," 8.
39. "Open Culture."
40. Galilei, "Sidereus Nuncius," 8.

It appears that the engravings in *Sidereus Nuncius* are a combination of actual observations of surface features and emphasized details[41] intended by the author to demonstrate the rough Earth-like topography of the Moon as opposed to the previously assumed perfectly smooth surface that might be expected from an object in the celestial region of the universe. It is also interesting to note that Galileo's engraving of the first-quarter Moon shows no lunar sea as would be expected. The field of view of the telescope was so small that it would be necessary to combine a number of sketches into one engraving and it is speculated that the sea called *Mare Serenitatis* just got overlooked.[42]

In *Sidereus Nuncius* Galileo describes the rugged surface of the Moon at length.[43] Although he made no such statement in the book, the rugged surface would imply then that if the Moon were similar to the Earth with a rocky surface, and it was known to be sailing through the celestial region of the universe, it would be possible for the Earth as a planet to be moving similarly through space as one might expect in a heliocentric universe.[44] This might mean that the Earth is not a fixed body at the center of the universe.

The Stars

In *Sidereus Nuncius* Galileo marveled at the great number of stars visible with his *perspicillum* that could not be seen with the naked eye. To demonstrate this fact, he even included engravings of two star fields as shown in Figure 4.7. Looking at the Milky Way he saw innumerable stars, clearing up prevailing disputes regarding the nature of the galaxy.[45] Further, Galileo noted that areas of milky brightness, like that of a cloud, are made up of clusters of stars.[46]

41. Del Santo et al., "Observing with Galileo's Telescope," 94.
42. Ibid., 97.
43. Carlos, *The Sidereal Messenger of Galileo Galilei*, 15–38.
44. Lindberg, "Galileo, the Church, and the Cosmos," 42.
45. Carlos, *The Sidereal Messenger of Galileo Galilei*, 42.
46. Ibid., 43.

All of these observations pointed to a reality that had not been expected or anticipated in the philosophical world of the early seventeenth century.

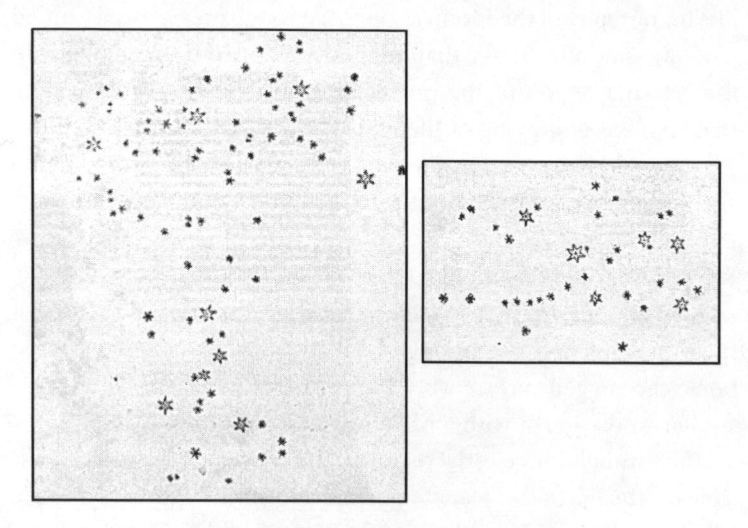

Figure 4.7. Star fields appearing in *Sidereus Nuncius*:[47] Orion on the left and Pleiades on the right.

The Satellites of Jupiter

One of the most significant discoveries made during Galileo's early observations was the existence of the satellites of Jupiter. He had already noticed the difference in the appearance of stars and planets in his *perspicillum*.[48] Planets appeared as small discs while the stars were point like in appearance. In *Sidereus Nuncius* he notes that early on the evening of January 7, 1610, he noticed three small stars near the planet, which he assumed were background stars, two on the west side and one on the east. When he looked at Jupiter the next night he saw all three on the west side. Clouds prevented any observations on January 9, but on January 10 he found

47. Galilei, "Sidereus Nuncius," 14–15.

48. Carlos, *The Sidereal Messenger of Galileo Galilei*, 46.

only two, both on the east side. He reasoned that the third was behind the disc of Jupiter. On January 11 after seeing two stars east of Jupiter he concluded that these three stars were moving about Jupiter just as Venus and Mercury appear to move about the Sun.[49]

Using our current knowledge of the orbits of the Jovian satellites it is possible to derive the positions of the satellites as they would have appeared in January 1609 and to compare those with the Galileo's drawings (see Figure 4.8). These show that his observations are a close match to reality and testify to Galileo's observational ability especially considering the somewhat primitive nature of his *perspicillum*. These observations did provide evidence of orbital motion about a central object other than the Earth, and Galileo did devote most of the remainder of his famous publication on the details of further observations of the what we now call the Galilean satellites of Jupiter.[50]

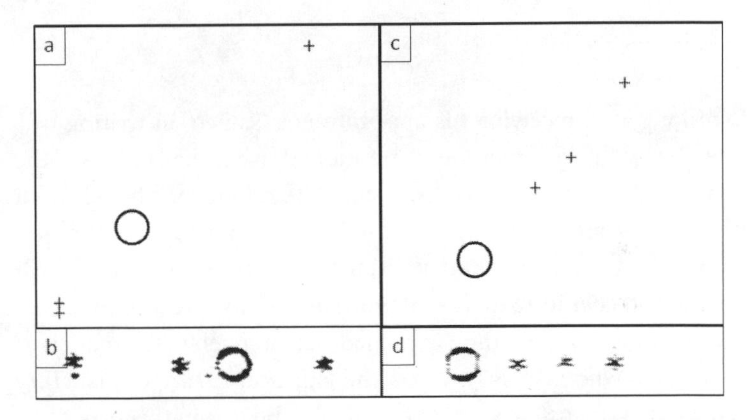

Figure 4.8. The Jovian satellites (a) January 7, 1609 derived positions as they would appear in the sky; (b) as drawn by Galileo; (c) 8 January 1609 derived position as they would have appeared in the sky; (d) As drawn by Galileo. (Graphics courtesy of Starry Night® Pro 7/ Simulation Curriculum Corp.)

Galileo was obviously impressed with this discovery. In the title page of *Sidereus Nuncius* and the dedication page (to Cosimo II

49. Ibid., 48.

50. Galilei, "Sidereus Nuncius," 17–27.

de' Medici) as well, he refers to the satellites as the Medicean stars after his former pupil. Following up on this dedication, Galileo applied for employment at the Tuscan court. He also requested at that time, "As to the title and scope of my duties, I wish in addition to the name of Mathematician his highness adjoin that of Philosopher . . . Whether I can and should have this title I should be able to show their highnesses whenever it is their pleasure to give me a chance to deal with this in their presence with the most esteemed men of their profession."[51] As a result on June 5, 1610, he was given the title "Chief Mathematician of the University of Pisa and Philosopher to the Grand Duke."[52] Obviously, following the discoveries made possible with his perspicillum, he wanted to investigate the natural philosophy as well as the mathematics of his world. This combination of the two separate disciplines in the person of Galileo would lead to the modern concept of science.

Saturn

Not long after receiving his appointment Galileo, in writing to a member of the Venetian court, mentioned that he had observed the planet Saturn to appear as three stars, the middle one being about three times larger than the ones on the side.[53] Such a configuration is shown in Figure 4.9. In that letter he requested that this be kept secret, and later in 1610 he sent word of this discovery to Johannes Kepler (1571–1630), the renowned astronomer/mathematician/astrologer who later discovered the elliptical nature of planetary orbits. As was customary at that time, he did this in the form of an anagram, which Kepler was unable to decipher, "s m a i s m r m i l m e p o e t a l e u m i b u n e n u g t t a u i r a s." Galileo eventually sent the solution, "*Altissimum planetam tergeminum observavi*," or "I have observed the highest planet tri-form."[54] Apparently he

51. Drake, *Galileo at Work*, 161.

52. Ibid.

53. Del Santo et al., "Galileo's Telescope," 53.

54. Van Helden et al., "The Galileo Project."

intended to publish this finding, but probably delayed doing so because he observed changes in the tri-form nature of its appearance which we now know to be due to the changing aspect of Saturn's rings. In 1612 he noted that the two small stars had actually disappeared, but he predicted their return in 1613.[55] It appears that he attributed this change of appearance to the changing aspect of the planet with respect to the Sun and the Earth, further evidence for a Copernican model of the solar system.[56] In August 1616, he wrote of a "new and strange phenomenon" in the appearance of Saturn, noting that two small globes were larger. They also appeared to be no longer round but in the shape of half an ellipse with dark triangles in the middle.[57] Our current knowledge of the changing aspect of Saturn's rings also makes it possible to reproduce the actual appearance of Saturn in the sky when Galileo made his observations (see Figure 4.9).

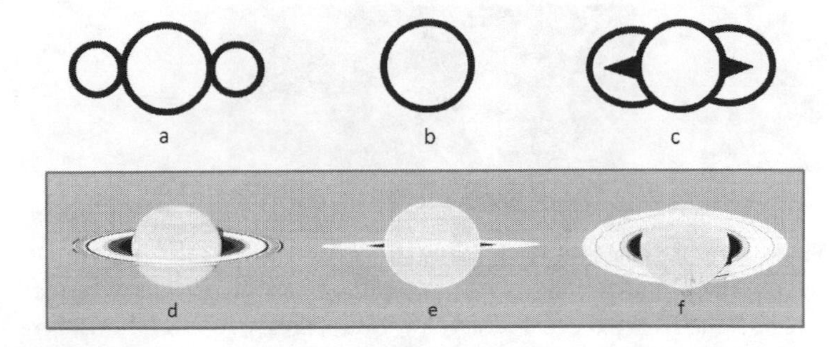

Figure 4.9. The appearance of Saturn as described by Galileo in 1610 (a), 1612 (b) and 1616(c) (illustrations by author) and as Saturn would appear in the sky in 1610(d), 1612(e) and 1616(f). (Graphics courtesy of Starry Night® Pro 7/ Simulation Curriculum Corp.)

55. Drake, *Galileo at Work*, 198.
56. Ibid., 212.
57. Ibid., 259–60.

Phases of Venus

Late in 1610 Galileo was able to observe the planet Venus.[58] He had not be able to do so earlier as the planet was too close to the Sun in the sky. When he did, he saw for the first time that it went through phases like the Moon. He documented these observations in his publication *Il Saggiatore.*[59] Again as with Saturn, we can recreate the appearance of Venus in late 1610 and early 1611 to compare with his engravings in *Il Saggiatore* (see Figure 4.10).

Figure 4.10. The phases of Venus 1610–1611 as drawn by Galileo (top)[60], and as recreated using Starry Night software (bottom). The dates of the recreated images from left to right are February 15, 1611, January 10, 1611, November 30, 1610, October 30, 1610, and July 15, 1610. (Graphics courtesy of Starry Night® Pro 7/ Simulation Curriculum Corp.)

The fact that Venus does go through these phases as seen from the Earth was interpreted by Galileo as evidence that Venus revolves about the Sun. He regarded that as proof of the Copernican universe, although that phenomenon is not incompatible with Tycho Brahe's theory.[61]

58. Ibid., 163.
59. Galilei, *Il Saggiatore*; cf. Galilei, "Il Saggiatore."
60. Galilei, "Il Saggiatore," 231.
61. Drake, *Galileo at Work*, 164.

Sunspots

In 1611 Galileo began a series of observations of the solar disk.[62] Because the Sun is much too bright to view directly through a telescope this was accomplished by viewing the Sun through light clouds or on a hazy horizon with his *perspicillum*. Projecting the image on to paper perpendicular to the length of the instrument was a method that was used as well. Spots on the Sun's disk had been observed previously using the naked eye looking through light clouds, but were largely dismissed as objects located between the Earth and the Sun, and not truly related to the Sun itself. In his work, *Istoria e dimostrazioni intorno alle macchie solari e loro accidenti* (History and demonstrations concerning sunspots and their properties),[63] Galileo presents his detailed observations. One such observation is reproduced in Figure 4.11 as an example of that detail. He goes on to argue from that extensive set of observations, that these spots are on the surface of the Sun and that their apparent motions are due to the fact that the Sun is rotating with a period of about one month. He even guessed that they appear to be dark because of their contrast with the bright solar disk.[64] The proposition that the Sun was subject to changes was at odds with the Aristotelian concept of an immutable celestial region and was the cause of a bitter dispute between Galileo and the Jesuit priest Christoph Scheiner (1573–1650).[65]

62. Ibid., 167.
63. Galilei, *Istoria E Dimostrazioni Intorno Alle Macchie Solari*, 59–96.
64. Del Santo et al., "Galileo's Telescope," 55–58.
65. Ibid.

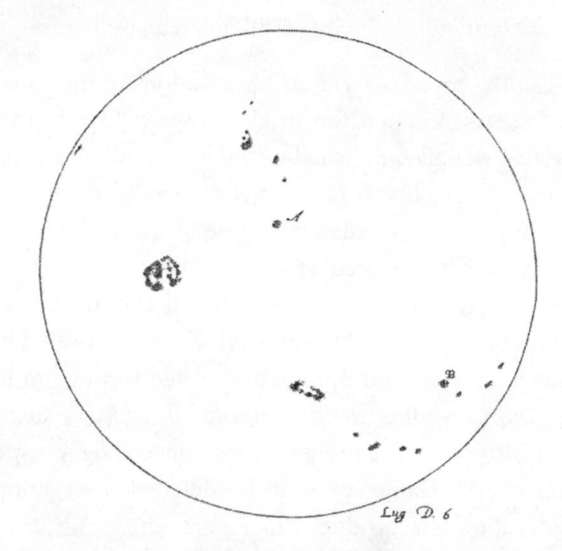

Figure 4.11. An example of Galileo's solar observations.[66]

Consequences of Galileo's Telescope

Galileo's telescope provided observations that obviously had profound consequences at the time it was constructed and revolutionized views of the world. Those observations showed that the Moon truly had a rocky surface; that the universe essentially had innumerable stars; that other planets had moons like the Earth; that those moons circled the planet just like the Earth might circle the Sun; that the planet Venus did circle the Sun rather than the Earth; and that changing blemishes on the Sun might indicate that the celestial region is not an immutable perfect realm. In short, the prevailing views of cosmology of the early seventeenth century were threatened. The idea that an appeal to authority could be regarded as a means to explain nature could no longer be accepted in the face of overwhelming evidence. The observations that were made with Galileo's telescopes eventually led to the combination

66. Galilei, *Istoria E Dimostrazioni Intorno Alle Macchie Solari*, 91.

of the disciplines of Renaissance mathematics and natural philosophy resulting in today's notion of science.

Bibliography

Bloom, Terrie F. "Borrowed Perceptions: Harriot's Maps of the Moon." *Journal for the History of Astronomy* 9 (1978) 117–22.

Carlos, Edward Stafford. *The Sidereal Messenger of Galileo Galilei: And a Part of the Preface to Kepler's Dioptrics Containing the Original Account of Galileo's Astronomical Discoveries*. London: Rivingtons, 1880.

Crombie, A. C. *Medieval and Early Modern Science*. 2 vols. Rev. ed. Garden City, NY: Doubleday, 1959.

Dear, Peter. *Revolutionizing the Sciences: European Knowledge and Its Ambitions, 1500–1700*. Princeton: Princeton University Press, 2001.

Del Santo, Paolo, Simone Esposito, Giorgio Strano and Maurizio Vannoni. "Observing with Galileo's Telescope." In *Galileo's Telescope, the Instrument That Changed the World*, edited by Giorgio Strano, 87–101. Milan: Giunti, 2011.

Del Santo, Paolo, Jim Morris, Rhoda Morris, Giorgio Strano and Albert Van Helden. "Galileo's Telescope." In *Galileo's Telescope, the Instrument That Changed the World*, edited by Giorgio Strano, 33–61. Milan: Giunti, 2011.

Drake, Stillman. *Galileo at Work: His Scientific Biography*. Chicago: University of Chicago Press, 1978.

Dreyer, J. L. E. *A History of Astronomy from Thales to Kepler, Formerly Titled History of the Planetary Systems from Thales to Kepler*. 2nd ed. New York: Dover, 1953.

Fahre, J. J., "The Scientific Works of Galileo (1564–1642) with Some Account of His Life and Trial Being a Review of Favaro's Edizione Nazionale delle Opere di Galileo (1890–1909)." Oxford: Clarendon, 1921. http://www.marcdatabase.com/~lemur/lemur.com/gallery-of-antiquarian-technology/philosophical-instruments/galileo-singer.

Galilei, Galileo. *Istoria E Dimostrazioni Intorno Alle Macchie Solari E Loro Accidenti*. Rome: Giacomo Mascardi, 1613.

———. *Il Saggiatore*. Rome: Giacomo Mascardi, 1623.

———. "Il Saggiatore." https://upload.wikimedia.org/wikipedia/commons/8/80/Works_of_Galileo_Galilei%2C_Part_3%2C_Volume_15%2C_Astronomy-_The_Assayer_WDL4184.pdf.

———. "Sidereus Nuncius." University of Oklahoma Libraries. https://digital.libraries.ou.edu/histsci/books/1466.pdf.

Grant, Edward. *The Foundations of Modern Science in the Middle Ages: Their Religious, Institutional, and Intellectual Contexts*. Cambridge History of Science. Cambridge: Cambridge University Press, 1996.

———. *A History of Natural Philosophy*. New York: Cambridge University Press, 2007.

———. *Science and Religion, 400 B.C. to A.D. 1550: From Aristotle to Copernicus*. Greenwood Guides to Science and Religion. Westport, CT: Greenwood, 2004.

Land, Barbara. *The Telescope Makers: From Galileo to the Space Age*. New York: Crowell, 1968.

Lindberg, David C. *The Beginnings of Western Science: The European Scientific Tradition in Philosophical, Religious, and Institutional Context, Prehistory to A.D. 1450*. 2nd ed. Chicago: University of Chicago Press, 2007.

———. "Galileo, the Church, and the Cosmos." In *When Science and Christianity Meet*, edited by David C. Lindberg and Ronald L. Numbers, 33–60. Chicago: University of Chicago Press, 2003.

Lindberg, David C., and Ronald L. Numbers. *When Science and Christianity Meet*. Chicago: University of Chicago Press, 2003.

Mando, Pier Andrea, Luca Mercatelli, Giuseppe Molesini, Maurizio Vannoni and Marco Verità. "The Quality of Galileo's Lenses." In *Galileo's Telescope, the Instrument That Changed the World*, edited by Giorgio Strano, 63–85. Milan: Giunti, 2011.

"Open Culture." http://www.openculture.com/2014/01/galileos-moon-drawings.html.

Pederson, Olaf. "Astronomy." In *Science in the Middle Ages*, edited by David C. Lindberg, 303–37. Chicago: University of Chicago Press, 1978.

Van Helden, Albert, Elizabeth S. Burr, Krist Bender, Adam C. Lasics, Adam J. Thornton, Martha A. Turner, Nell Warnes, Lisa Spiro, and Megan Wilde. "The Galileo Project." http://galileo.rice.edu/about.html.

5

Galileo's Contribution to Mechanics

Asim Gangopadhyaya

We generally remember Galileo for the fantastic advances he made in observational astronomy and his brilliant defense of the heliocentric theory of our solar system. He also made enormous contributions to the foundation of mechanics which are often overlooked.

In today's language, Galileo would be called a theoretical as well as an experimental physicist. In addition to carefully performed experiments, Galileo is also known as a master of *Gedanken* experiments (thought experiments). While we will discuss some of his thought experiments in this chapter, the main emphasis will be on the experiments he conducted and his theoretical deductions that profoundly influenced Newton. In a recent translation of *Principia*, the authors claim that Newton gave credit to Galileo for both the first and the second laws of motion that form much of the basis for classical mechanics.[1]

In this chapter I will review his contribution to mechanics including his work on pendulums, the Galilean Theory of Relativity—the precursor to Einstein's Theory of Relativity, investigations of uniformly accelerated objects including freely falling bodies, objects on frictionless, inclined planes, as well as his adventures with the "New Machine." For an extensive description of his wide

1. Cohen and Whitman, *The Principia*.

ranging work in this area, we refer the readers to the work by Roberto Vergara Caffarelli.[2]

Galileo and Mechanics

Experimental investigations of mechanical systems involve a careful study of the time evolutions of positions, i.e., the recording of a sequence of positions parameterized by increasing values of time. All such measurements require collection of accurate data for both position and time. While measurement of positions can be done very accurately and consistently, the same cannot be said about the measurements of temporal intervals. The sundial and its variants have been used since antiquity to measure time, but these are of value only when the sun is visible. Hour glasses that use the flow of a fixed amount of sand and water clocks were great advancements in time measurement due to their portability and because they could be used at night. However, the measurement of time was revolutionized when oscillatory systems came into use.

Measurement of Time

Use of oscillatory mechanics to measure time intervals was in existence in Europe before Galileo's time. However, many such measurements, as far as we know, were made using a trial and error method. During his student days in Pisa (1583–85), Galileo is known to have carried out the first controlled scientific study of a simple pendulum, a device consisting of a ball hanging by a flexible string. One variable that affects the period of a pendulum is its length: the period is proportional to the square root of the length l ($\tau \propto \sqrt{l}$) when all other variables are kept fixed. Some authors have argued that Galileo did not discover this functional dependence till much later in his life since he did not write about it before his *Discorsi*. But that is hard to believe considering the extensive analysis Galileo had done on the isochronicity of pendulums; i.e., the

2. Vergara Caffarelli, *Galileo Galilei and Motion*.

equality of frequencies for all pendulums of the same length.[3] He also found that two simple pendulums of the same length, but with different suspended masses (of the same size), will have exactly the same period, but the one with smaller mass will die out earlier, a phenomenon we can easily explain today using Newton's laws.[4] It would be almost another one hundred years before Christian Huygens would derive the dependence of the time period τ on l and g (acceleration due to gravity $\approx 9.81\text{m/s}^2$): $\tau = \frac{1}{2\pi}\sqrt{\frac{l}{g}}$ and build a clock based on the compound pendulum.

Galilean Relativity

One of the basic tenets of relativity is that the laws of physics appear to be the same for all inertial observers[5]—observers that are moving with constant velocities[6] with respect to each other and with respect to distant stars. For practical purposes, the frames that are moving with constant velocity with respect to the surface of Earth can be considered to be inertial frames. As we all know, we can drink a cup of coffee just as easily on a plane or a train that is coasting with a constant velocity as standing at a station—all three are examples of inertial frames. However, drinking the same coffee in an accelerated system such as an accelerating car or a roller coaster is a very different proposition. In 1632, Galileo introduced this principle of relativity; i.e., the equivalence of all inertial frames, in order to argue that the motion of the Earth around the Sun, or the Earth's spin about its axis, should make no more discernible difference on the free falls of terrestrial bodies than what would be expected on a stationary Earth. This was, of course, to

3. Ibid.

4. Two balls with same size and speed will experience exactly the same resistive force due to air. From Newton's second law, this implies that the less massive ball will go larger deceleration and hence will slow down faster.

5. Also called inertial frames of reference.

6. By velocity, we mean speed and the direction of motion. An object moving with a constant velocity travels along a straight line with constant speed.

support the Copernican heliocentric system. In particular, Galileo introduced a thought experiment in *Dialogues Concerning the Two Chief World Systems* where he surmised that a ball dropped from the top of the mast of a ship will hit the ship at the base of the mast independent of whether the ship was stationary or moving with a constant velocity. A uniform motion of the ship will not cause the ball to fall behind the mast. His claim was not without basis. He had seen the independence of the vertical and horizontal motions of a projectile and explained that the uniform horizontal motion of the ship should have no effect on the falling ball. In other words, an experiment carried out below the deck of a cruising ship would not be able to determine whether the ship was stationary or moving with a constant velocity. This claim was apparently substantiated by the observations of his friend, Giovanni Francesco Sagredo, a mathematician and an avid traveler, and experimentally verified by an empiricist and mathematician, Pierre Gassendi.[7]

No one before Galileo had stated this principle of relativity as clearly. The impact of this principle on the developments of mechanics, electrodynamics, and special relativity cannot be exaggerated.

Freely Falling Bodies

When things move under the force of gravity alone, the resulting motion is called free fall. This is probably one area of mechanics where Galileo's contributions are well known. His observations of free fall formed the basis of Newton's and Einstein's laws of gravity. He supposedly dropped two objects of different mass from the top of the tower of Pisa, and they both landed at the same time.[8] From this, he observed that all masses accelerate at the same rate under gravity. Similar experiments are commonly performed in classrooms today, leading to exactly the same result every time. This observation regarding free fall violated Aristotle's conjecture

7. Gassendi, *De Motu*.

8. It is widely believed that this was just another thought experiment and Galileo never really carried out this experiment.

that the velocities of such falling objects are proportional to their masses,[9] i.e., an object twice as heavy as another should move with twice the velocity and should reach Earth in half the time.

Galileo buttressed his experimental observation with one of the best thought experiments in the history of physics. To understand his arguments let us assume that Aristotle was right, i.e., a lighter object did come down slower than a heavier object and took a longer time to reach the earth than a heavier object starting from the same height. If we tie these two objects together and let them fall, the heavier one will be pulling down on the lighter one and hence speeding it up, and at the same time the lighter one will be pulling the heavier one up and thus slowing it down. The combination should then come down with a common velocity which would be between those of the lighter and the heavier objects had they fallen independently. But then the combination should have a higher weight than both objects, and by Aristotelian principle, it should come down with even greater velocity than either one of them. Thus, we reach a conundrum. The only way to solve it is that each would come down with the same velocity and acceleration, independent of their masses.[10]

According to Newton's second law of motion, the acceleration depends on the ratio of force to the inertia of the body, called the "inertial mass"; i.e., $a = \frac{F}{m_i}$. The gravitational force arises due to interaction between the object and the Earth, and is proportional to m_G, another inherent attribute of the object that we call the "gravitational mass," i.e., $F = m_G g$. The acceleration for an object in free fall is then given $a = \frac{F}{m_i} = (\frac{m_G}{m_i})g$. For this acceleration a to be independent of the mass requires that $m_G = m_i$. This intriguing equality is the basis for Newton's law of gravity and Einstein's principle of equivalence, the starting point for the General Theory of Relativity.

9. Heavier objects do fall faster than lighter objects of the same size due to air resistance.

10. In fact we often see bodies falling at different rates. This is due to air resistance (friction), and it was Galileo's genius to remove that variable.

Accelerated Linear Motion

In the last section, we saw the foundational influence Galileo's observations of free fall had on the work of Newton and Einstein. However, vertical free fall was too fast to measure with the technologies of the sixteenth century. Accurate recording of the time and positions of a falling object was virtually impossible. So, he needed to slow down the motion of the objects, and he did that in two ingenious ways:

1. by letting objects slide on inclined planes;
2. by using a newly invented machine that partially balanced the weight of a falling object using a counter weight.

First we discuss Galileo's use of inclined planes to reduce the accelerations of objects. Inclined planes allowed Galileo to carry out careful measurement of the positions of accelerating objects as functions of time. One of the major observations he made was that the distance traveled in time t along an inclined plane increased as the square of the elapsed time, i.e., $d \propto t^2$. This implied that the speed of the object increased at a constant rate along the plane, and that this rate depended on the inclination. From these experiments, and through theoretical reasoning, he derived the law that governs the motion of objects on inclined planes. We know that if we place a ball on a horizontal plane, it does not move at all. This is because the weight of the ball is supported by an equal but opposite force exerted by the plane. What would happen if the plane were inclined? Galileo answers this through the following argument. On an inclined plane the gravitas of the ball would be *partially* balanced by a force perpendicular to the plane. What force parallel to the plane would we then need to insure that the ball does not roll? He argued that for the ball not to move the moments of the weight (what we call "torque") and the balancing force[11] must cancel each other out. He then showed that this required that the ratio of the balancing force to the gravitas be the same as the height

11. It is important to note that the history of moments of forces goes back to Archimedes.

of the inclined plane to its length. In other words, the balancing force needs to be equal to (*height/length*)×*weight*. The height and length are shown in Figure 5.1. In today's language this would be written as $F = W \times \sin \vartheta$

Figure 5.1: A block sliding on an inclined plane of height H and length L.

where the angle ϑ is the angle of inclination of the plane, $\sin \vartheta$ = (*height/length*). From this he deduced that the distances traveled by an object on planes of varying inclinations within a fixed time interval Δt were proportional to the ratios of their heights to lengths. In fact, he was flirting with Newton's second law.

From Newton's laws we know that the distance traveled on an inclined plane is given by

$$d = \tfrac{1}{2}a(\Delta t)^2 = \tfrac{1}{2}g \sin \vartheta \, (\Delta t)^2$$

Thus, the ratios of distances traveled in time Δt on planes of varying inclinations would be given by

$$\frac{d_1}{d_2} = \frac{\tfrac{1}{2}g \sin \vartheta_1 (\Delta t)^2}{\tfrac{1}{2}g \sin \vartheta_2 (\Delta t)^2} = \frac{\sin \vartheta_1}{\sin \vartheta_2} = \frac{H_1/L_1}{H_2/L_2}$$

This is exactly what we get from the application of Newton's laws.

Two-dimensional Motion of a Projectile

As the following diagram shows (Figure 5.2), Galileo launched projectiles with different initial speeds from a horizontal launcher into a two-dimensional vertical plane. As the starting speed increased, so did the range of the projectile,[12] and they were linearly proportional to each other ($R \propto v$). Galileo was among the first to discover one of the most important aspects of two-dimensional motion: the motion along the horizontal direction was completely independent of the motion along the vertical direction. He also noted that the uniform velocity along the horizontal direction and free-fall along the vertical resulted in a parabolic trajectory.

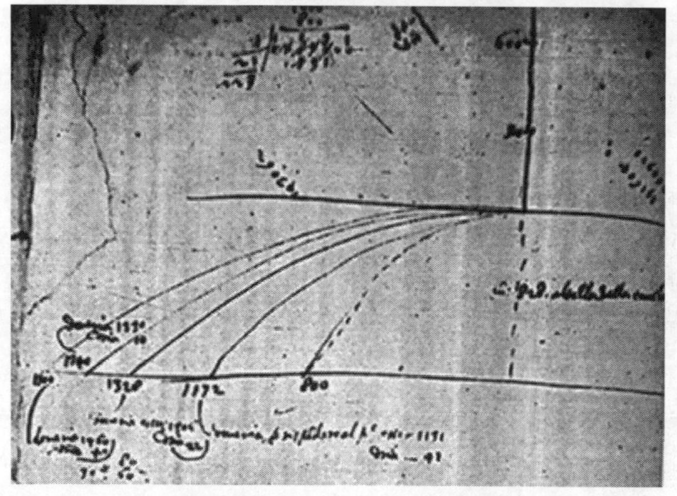

Figure 5.2: Trajectories of projectiles for varying starting velocities (Galilean manuscripts, 1608).

Let us present here another example in which two independent motions are superimposed on each other. In this ingenious experiment, a ball is released off center on an inclined trough. If the trough is kept horizontal the ball executes oscillatory motion with a fixed period determined by the curvature of the trough. When the ball is released on an inclined trough (see Figure 5.3), it comes down along the axis of the trough with an increasing velocity and a transverse oscillation. When this experiment is carried out in a laboratory, one finds that the extreme rightward point of each

12. I.e., how far these projectiles landed from the point of projection along a horizontal axis denoted by numbers such as 800, 1172 and 1320 units of length.

oscillation provides us positions along the length of the trough at equal intervals, and the distances of these points[13] from the starting position increases in ratios of squares of natural numbers, i.e. $d_n = n^2 d_1$. Thus, we see that motion along the axis of the trough obeys the laws of an inclined plane; it is completely independent of transverse oscillations. Even though this experiment might have been performed by Galileo, we do not find any mention of it in the literature.[14]

Figure 5.3: Combination of a linear motion on an inclined plane and a transverse oscillatory motion. Courtesy of Harvard Natural Sciences Lecture Demonstrations and Prof. Owen Gingerich. For more details, see http://sciencedemonstrations.fas.harvard.edu/presentations/scantling-and-ball.

13. The distance of the n-th peak on the right from the starting position is denoted by d_n.

14. Communications with Profs. Roberto Vergara Caffarelli and Paolo Palmieri.

Newton's Laws of Motion: Inertia

According to Aristotle, to keep an object moving even at a constant velocity, we need an external force on the object. Galileo put this principle to the test. In Figure 5.4, a ball released from point A rises to point B on a second plank. If the inclination of the plank were lowered, it would rise to C. Each time it goes up to the same height as the starting point A. If we keep on lowering the inclination of the second plank, it will travel farther and farther to reach the same height. What if we make the second plank horizontal? In this case the ball should keep going farther and farther in its futile effort to reach the height of A—this is the law of inertia. That is, we do not need to keep pushing an object to keep it moving with constant velocity on a frictionless horizontal surface. This is the statement of Newton's First law which Newton himself attributed to Galileo.

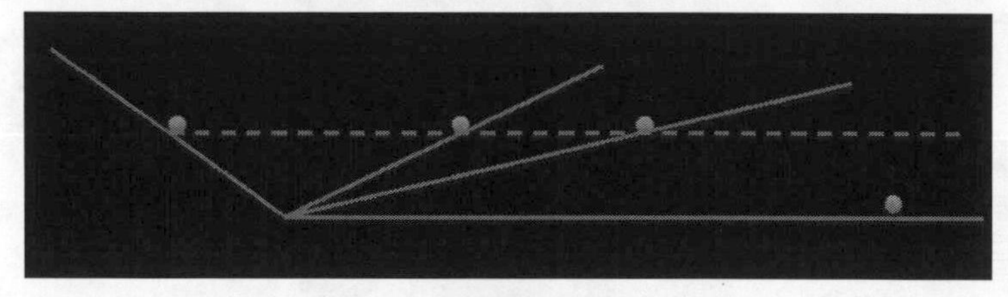

Figure 5.4

Galileo's New Machine

We now describe an experiment that was carried out by Galileo in his later years which brings us very close to the laws of dynamics as we know them today. A similar piece of equipment was built by George Atwood about 150 years later and is now known as the Atwood Machine. We have schematically shown the apparatus in Figure 5.5. Two masses, m_1 and m_2, are supported by pulleys. Caffarelli's reconstruction of the equipment shows one pulley; we have chosen to use two pulleys to ensure that the pulleys could be made small and have relatively little effect on the vertical motions of the weights.

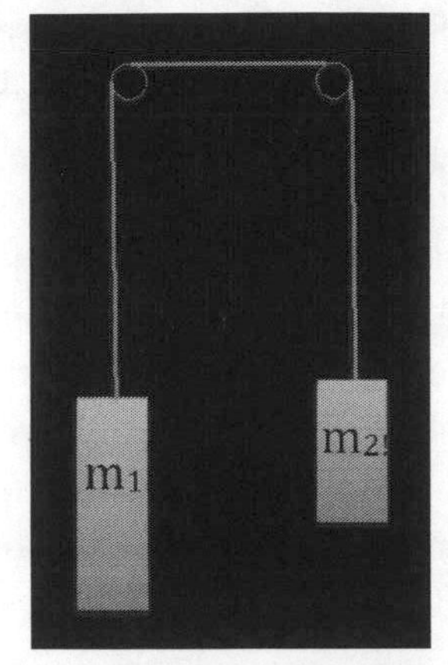

Figure 5.5: Galileo's New Machine

Galileo experimented with different relative values of these masses. Choosing equal masses, he demonstrated that one need not keep pushing these objects for them to have uniform velocities - a statement of the law of inertia, but this time for vertical linear motion. He also carried out quantitative studies of different masses. (This may well be why Newton gave him credit for the second law of motion, $F = ma$.) Very importantly, Galileo carried out an interesting experiment on this apparatus that would yield the principle of momentum conservation. He kept one mass (say m_2) on a support and lifted the other mass m_1 to a certain height h and slackened the rope linking them together. When he the released the mass m_1, it came down with increasing speed. Eventually the rope tightened and the second mass started moving in what he called a "violent motion." Galileo discovered that the velocity v_1 of m_1 right before the rope tightens bears a relationship to the common speed V of the combination just after the rope tightens: $m_1 v_1 = (m_1 + m_2)V$. This is actually the statement of momentum conservation for a perfectly inelastic collision, a corollary of Newton's second and third laws of motion.

Concluding Remarks

We use the phrase "Renaissance person" to refer to an individual who excels in many areas at the same time. Galileo was a figure not only historically from the late European Renaissance, but also was one whose mind and its achievements touched many areas: physics, astronomy, mechanics, religion, and politics. Although often overshadowed by the controversies he was a part of, his contributions to the understanding of motion, inertia, acceleration, and gravity are remarkable. Explaining both his real and his thought experiments in mechanics gives us yet another way to appreciate his lasting contributions to physics and humanity.

Acknowledgment

This article would not have been completed if not for the immense help I received from my colleagues, Jeffry V. Mallow and Thomas T. Ruubel. I am very grateful to them.

Bibliography

Gassendi, Pierre. *De Motu Impresso a Motore Translato*. (1642) New World Encyclopedia. http://www.newworldencyclopedia.org/entry/Pierre_Gassendi.

Newton, Isaac. *The Principia: Mathematical Principles of Natural Philosophy*. Translated by I. Bernard Cohen and Anne Whitman, assisted by Julia Budenz. Berkeley: University of California Press, 1999.

Vergara Caffarelli, Roberto. *Galileo Galilei and Motion: A Reconstruction of 50 Years of Experiments and Discoveries*. Berlin: Springer, 2009.